BASIC ROBOT BUILDING
with LEGO Mindstorms NXT 2.0

John Baichtal

800 East 96th Street,
Indianapolis, Indiana 46240 USA

Basic Robot Building with LEGO Mindstorms NXT 2.0

Copyright © 2013 by John Baichtal

ISBN-10: 0-7897-5019-8

ISBN-13: 978-0-7897-5019-8

Library of Congress Cataloging-in-Publication data is on file.

Printed in the United States of America

First Printing: December 2012

Trademarks

All terms mentioned in this book that are known to be trademarks or service marks have been appropriately capitalized. Que Publishing cannot attest to the accuracy of this information. Use of a term in this book should not be regarded as affecting the validity of any trademark or service mark.

Warning and Disclaimer

Every effort has been made to make this book as complete and as accurate as possible, but no warranty or fitness is implied. The information provided is on an "as is" basis. The authors and the publisher shall have neither liability nor responsibility to any person or entity with respect to any loss or damages arising from the information contained in this book.

Bulk Sales

Que Publishing offers excellent discounts on this book when ordered in quantity for bulk purchases or special sales. For more information, please contact

U.S. Corporate and Government Sales
1-800-382-3419
corpsales@pearsontechgroup.com

For sales outside of the U.S., please contact

International Sales
international@pearsoned.com

Editor-in-Chief
Greg Wiegand

Executive Editor
Rick Kughen

Development Editor
Rick Kughen

Managing Editor
Sandra Schroeder

Project Editor
Seth Kerney

Copy Editor
Chuck Hutchinson

Indexer
Ken Johnson

Proofreader
Paula Lowell

Technical Editor
James F. Kelley

Publishing Coordinators
Cindy Teeters
Romny French

Interior Designer
Mark Shirar

Cover Designer
Anne Jones

Compositor
Trina Wurst

Contents at a Glance

Introduction   1

CHAPTER 1   Unboxing the LEGO Mindstorms NXT Set   7

CHAPTER 2   Project: Backscratcher Bot   37

CHAPTER 3   Anatomy of the NXT Brick   57

CHAPTER 4   Introduction to Programming   81

CHAPTER 5   Project: Clothesline Cruiser   95

CHAPTER 6   Building Stronger Models   123

CHAPTER 7   Know Your Sensors   131

CHAPTER 8   Advanced Programming   143

CHAPTER 9   Project: Rebounder   157

CHAPTER 10 Expanding on Mindstorms   181

Glossary   195

Index   201

Table of Contents

Introduction   1

Conventions Used in This Book   4

Special Elements   4

Chapter 1   Unboxing the LEGO Mindstorms NXT Set   7

The Box   7

Under the Flap   8

Opening the Box   9

The Contents   9

Reading Material   10

Connector Pegs   11

3M Connector Pegs   12

Connector Pegs with Bushing   13

Connector Pegs with Cross-Axle Ends   14

2M Axle Peg   14

Connector Peg with Towball   15

1/2 Connector Pegs   16

Bushings   16

Cross Axles   17

Cross Axles with End Stops   18

Wheels and Treads   19

Gears   20

Teeth   21

Balls   21

Beams With Pegs   22

Beams   23

Angle Beams   24

Car Parts   25

Steering Links   26

Angle Elements   26

Cross-Axle Connectors   27

Cross Blocks   27

More Miscellaneous Parts   28

Peg Joiner   29

Motors, Wires, and Sensors   31

USB Cable   34

Mindstorms Wires   35

The NXT Brick   35

Next Chapter   36

Chapter 2   Project: Backscratcher Bot   37

Adding Batteries to the NXT Brick   38

Parts You Need   40

Step-by-Step Assembly Instructions   42

Programming the Backscratcher Bot   49

Running the Backscratcher Bot   56

Next Chapter   56

Chapter 3   Anatomy of the NXT Brick   57

The Brick   58

Buttons   59

Ports   59

Connector Holes   61

Reset Button   62

Menus   63

My Files   63

NXT Program   66

Try Me   67

The View Menu   68

NXT Datalog   70

Settings   70

Bluetooth   73

Powering Your NXT   77

Resetting a Crash   78

Updating NXT Firmware   78

The Next Chapter   79

Chapter 4   Introduction to Programming   81

System Requirements   82

Installing the Software   83

Installing on a PC   83

Installing on a Mac   84

NXT-G 101   85

The Programming Block   85

Commonplace Blocks   87

Programming the Backscratcher Bot   92

Create the Program   92

Connect to the NXT Brick   93

Download the Program   94

The Next Chapter   94

Chapter 5   Project: Clothesline Cruiser   95

Parts You Need   96

Step-by-Step Instructions   97

Programming the Clothesline Cruiser   119

Setting Up the Clothesline   120

What to Do With Your Cruiser?   121

The Next Chapter   122

Chapter 6   Building Stronger Models   123

Use Multiple Pegs   123

Connect Each Part to as Many Others as Possible   123

Reinforce Corners with Angle Beams   124

Use Combination Parts and Cross Blocks   124

Attach Cross Axles   125

Combine Technic and System Bricks   126

Use Chassis Bricks   128

The Next Chapter   129

Chapter 7   Know Your Sensors   131

Mindstorms Sensors   131

Touch Sensors   132

Ultrasonic Sensors   133

Color Sensors   133

Sound Sensor   134

Motor   135

Calibrating Sensors   136

Third-Party Sensors   137

    Compass Sensor   138

    Passive Infrared (PIR) Sensor   138

    Wi-Fi Sensor   138

    Magnetic Sensor   139

    Flex Sensor   139

    Voltage Sensor   140

    Barometric Sensor   140

    Inertial Motion Sensor   140

The Next Chapter   141

Chapter 8   Advanced Programming   143

Data Wires   144

    Green Wires   144

    Yellow Wires   145

    Orange Wires   145

    Gray Wires   146

Connecting Wires   147

Additional Blocks   148

    Variable Block   148

    Constant Block   149

    Random Block   150

    Keep Alive Block   150

    Light Sensor Block   151

    Rotation Sensor Block   151

    Display Block   152

    Bluetooth Block   152

    Logic Block   153

Creating Your Own Blocks   154

The Next Chapter   156

Chapter 9   Project: Rebounder   157

Parts You Need   158

Step-by-Step Instructions   159

A Note About Tank Treads   175

Programming the Rebounder   176

The Next Chapter   180

Chapter 10 Expanding on Mindstorms   181

  Read Blogs   181

    The NXT STEP   181

    Mindstorms   182

  Design Virtual Models   183

  Attend Gatherings   184

    LUGs   185

    Conventions   185

  FIRST LEGO League   186

  Read *BrickJournal*   188

  Expand Your Collection   188

    Bricklink   189

    Pick a Brick   189

    LEGO Education   189

  Third-Party Brick Makers   190

    Omni Wheels   190

    Bricktronics   190

    Tetrix   191

  Print Your Own   192

Glossary   195

Index   201

About the Author

John Baichtal is a contributor to *MAKE* magazine and Wired's GeekDad blog. He is the co-author of *The Cult of LEGO* (No Starch) and author of *Hack This: 24 Incredible Hackerspace Projects from the DIY Movement* (Que). Most recently he wrote *Make: Lego and Arduino Projects for MAKE*, collaborating with Adam Wolf and Matthew Beckler. He lives in Minneapolis, Minnesota, with his wife and three children.

Dedication

This book is dedicated to my lovely wife Elise and my LEGO-obsessed kids. Sorry about all the robots cluttering up the house!

Acknowledgments

I'd like to thank my editor, Rick Kughen, and my frequent collaborator, Adam Wolf, for their help with this book.

We Want to Hear from You!

As the reader of this book, *you* are our most important critic and commentator. We value your opinion and want to know what we're doing right, what we could do better, what areas you'd like to see us publish in, and any other words of wisdom you're willing to pass our way.

We welcome your comments. You can email or write to let us know what you did or didn't like about this book—as well as what we can do to make our books better.

Please note that we cannot help you with technical problems related to the topic of this book.

When you write, please be sure to include this book's title and author as well as your name and email address. We will carefully review your comments and share them with the author and editors who worked on the book.

Email: feedback@quepublishing.com

Mail: Que Publishing
            ATTN: Reader Feedback
            800 East 96th Street
            Indianapolis, IN 46240 USA

Reader Services

Visit our website and register this book at quepublishing.com/register for convenient access to any updates, downloads, or errata that might be available for this book.

Introduction

You're holding the most amazing building set in the world. Now what?

It can be a little intimidating, when faced with all those possibilities. Mindstorms has built up such a massive following that a veritable ecosystem has developed—modelmakers sharing their design files, programmers creating new blocks, and conventions gathering together builders from around the world. What could you do with all of that? The easy answer is, a lot. The more challenging question is, where do you begin in such a vast pool of knowledge? The aim of this book is to simplify the experience and make it easy and fun.

Basic Robot Building with LEGO Mindstorms NXT 2.0 shows you how to build three easy models, using only the parts found in a LEGO Mindstorms NXT 2.0 boxed set. Everything you need is in this book and your set!

How can we do projects with Mindstorms without knowing what's in the set? In Chapter 1, "Unboxing the LEGO Mindstorms NXT Set," I break down everything you get as we unbox the entire set.

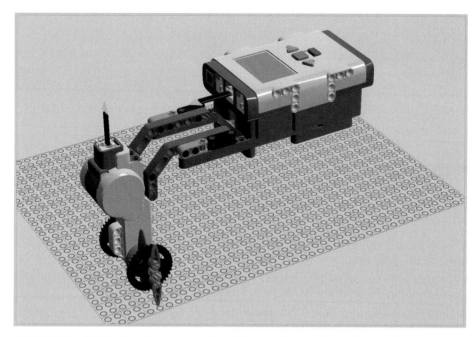

FIGURE I.1   Build the Backscratcher Bot and cure itchy backs forever!

In Chapter 2, "Project: Backscratcher Bot," we dive in the deep end of the LEGO pool as we build our first robot, a portable Backscratcher Bot (see Figure I.1) that needs no programming know-how to get working.

The most important element in the set is the NXT brick, a clever miniature computer that controls the motors and sensors and helps you turn a pile of plastic bricks into a robot. In Chapter 3, "Anatomy of the NXT Brick," we'll delve into the NXT and examine all of its capabilities and menus.

To reach the full potential of the NXT, however, we'll have to learn how to program it. In Chapter 4, "Introduction to Programming," we'll examine NXT-G, the Mindstorms programming environment, and write our first program.

Then it's time for our next robot! Chapter 5, "Project: Clothesline Cruiser," provides instructions on building the Clothesline Cruiser (see Figure I.2), a robot that travels via rope! We'll also program the robot to control its movements.

FIGURE I.2   The Clothesline Cruiser transports itself along a length of clothesline.

After we've got a couple of robots under our belts, we'll delve deeper into the mysteries of Mindstorms! Chapter 6, "Building Stronger Models," introduces you to advanced building techniques, offering tips on building stronger and smarter models.

Chapter 7, "Know Your Sensors," immerses you in the world of sensors, those electronic gadgets that plug into the NXT brick and send it data. You can use sensors to do everything from judging distance to detecting an object's color. Let's find out how they work and what they can do!

Chapter 8, "Advanced Programming," offers advanced programming techniques, introducing you to a bevy of concepts to help you make your robots even cooler!

We get to put these techniques to the test in Chapter 9, "Project: Rebounder," in which we build our third and final robot. The Rebounder (see Figure I.3) is miniature tank-treaded robot that rolls around blissfully, then rebounds when it bumps into something.

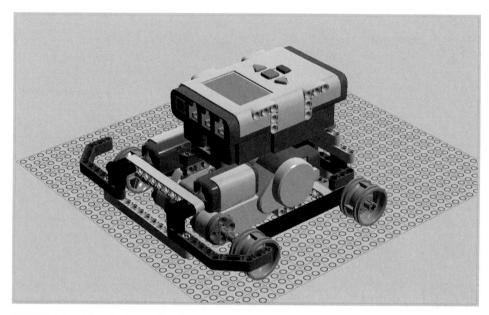

FIGURE 1.3   The Rebounder is an autonomous robot that reacts to walls by moving away from them.

Finally, Chapter 10, "Expanding on Mindstorms," suggests some next steps for advancing your LEGO knowledge, as well as expanding your supply of parts. The Mindstorms ecosystem is vast, and I'll show you how to smartly expand your capabilities.

There you have it. Mastering LEGO Mindstorms is a rewarding journey that teaches you about everything from mechanical engineering to computer programming. Let's get started!

Conventions Used in This Book

I think this book should be fairly easy to figure out without much explanation from me, although there are a few teaching tools I use along the way that deserve some discussion here.

Special Elements

This book includes a few special elements that provide additional information not included in the basic text. These elements are designed to supplement the text to help you get up to speed with your LEGO Mindstorms NXT 2.0 quickly and easily.

TIP

A *tip* is a piece of advice that helps you accomplish a task—whether it be a hands-on building task or a programming chore—easily and with little to no heartache.

NOTE

A *note* is designed to provide information that is useful and/or interesting but not crucial for the task at hand. Notes often contain ancillary information that is good to know, but won't be crippling if you skip it.

CAUTION

A *caution* is the publishing equivalent of a big, red stop sign. When you see one of these, read it and do what it says! Failure to do so can have a variety of repercussions—all the way from a Mindstorms bot that doesn't work, to a potentially dangerous situation.

SIDEBAR

Some Extra Thoughts

Sidebars are where I've tucked information that is relevant, but slightly off-task or non-essential, such as juicy bits of background and behind-the-scenes info on LEGO Mindstorms.

Unboxing the LEGO Mindstorms NXT Set

When you drooled over product photos on the Mindstorms box or dug through the parts inside, no doubt you found the breadth and complexity of the parts a little unnerving. Looking at all those LEGO beams and connectors can be a little intimidating, not knowing what they're for or how they fit together. In this chapter, we look at every part that comes with the set and talk a little about each one.

The Box

The LEGO Mindstorms NXT 2.0 box (see Figure 1.1) intrigues with all the cool possibilities you can imagine are inside, not the least of which are the great—albeit complex—models LEGO shows you how to make with the set. The headliner is Alpha Rex, the humanoid robot in the center of the box. There's a little science fiction in the rendering; the color sensor doesn't actually emit a visible beam of light, and the eyes, which are the ultrasonic sensor, don't actually light up.

A little creative license was used here...

FIGURE 1.1 The LEGO Mindstorms NXT 2.0 box.

Still, you can actually build the Alpha Rex, following the instructions contained in the Mindstorms software that came with the set.

NOTE

Alpha Rex Directions

I didn't include directions for building Alpha Rex in this book because instructions for building this bot are included with the set. Instead, I focused on building three bots that aren't detailed in the instructions. That means between this book and the instructions that came with the set, you now have details on how to build *seven different bots*!

Under the Flap

LEGO is really good about dressing up the box with tantalizing glimpses of the contents, all the better to get customers to reach for their wallet (see Figure 1.2). Looking at all those beautiful parts and robots displayed, how could you not buy the set?

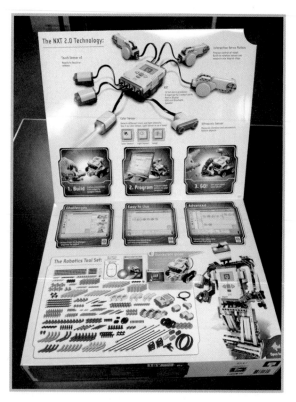

FIGURE 1.2   This is 100% pure geeky goodness.

This great artwork is also educational, explaining the difference between the sensors and giving a brief intro on how programming works. What I found most exciting is the huge spread showing all the components that come with the set.

Opening the Box

Cracking open the box, you see a bunch of plastic bags of LEGO parts stuffed into the box (see Figure 1.3). You'll very quickly realize the Mindstorms box is probably not a container that you can use long term for storing your set, particularly if you use it a lot. While pretty, it's merely a cardboard box with no dividers or reinforcement, so it will get squashed very quickly.

FIGURE 1.3 From "wow" to "ho-hum," the inside of the Mindstorms box is, well, just a box.

The Contents

Next, let's dump out everything inside the box (see Figure 1.4). Look at all this cool stuff! The plastic bags hold the LEGO beams and connectors, and the white cardboard boxes hold the NXT brick and the sensors, motors, and wires. When you look at the parts like this, it doesn't seem like a $280 set, but when you start building cool robots, you'll think differently.

FIGURE 1.4 Mindstorms parts—just add imagination.

WHAT YOU DON'T GET WITH THE SET

There is a lot of cool stuff that LEGO has not seen fit to include in the set. For instance, there is a huge variety of gears available that you'll never find in the basic Mindstorms set. You might also notice that the parts found in the set are fairly blah when it comes to color. Be sure to read Chapter 10, "Expanding on Mindstorms," in which I discuss where to find all sorts of parts to expand upon Mindstorms' possibilities.

Reading Material

Let's dive in! First, let's begin with the printed stuff that comes with the set, as shown in Figure 1.5. The Test Pad, on the left, is a poster-sized track that you can use with a rolling robot. It has a circuit that you can use to test a line-following robot, as well as color swatches to test its color sensor. The stickers are just for looks, obviously, and are used to dress up the four default models that LEGO suggests you make with the set. The *LEGO Mindstorms User Guide* offers basic instructions on how to build robots and program them, and the disc holds the software you need to do so. We talk more about the disc and its programs in Chapter 4, "Introduction to Programming."

FIGURE 1.5   The user guide and other materials included with the set.

Connector Pegs

In many ways, these little tubes—*connector pegs* in LEGO parlance—are the glue that holds your robot together (see Figure 1.6). The pegs connect two LEGO elements via their holes, and you'll use them all the time. Interestingly, the black and gray pegs are nearly identical and in many cases can be used interchangeably. The critical difference between the two is that the black pegs have little ridges on them that add friction, to reduce the tendency of a single peg to rotate on its own. By contrast, the gray pegs are smooth and can be used as axles. You get 88 black pegs with your set and 6 gray ones.

> **NOTE**
>
> **Your Mileage May Vary**
>
> You might get a different number of parts than what I detail here. For example, I got seven gray connector pegs in my set. Even LEGO's vaunted quality control messes up sometimes!

Gray pegs are smooth and can be used as axles

Black pegs have ridges that deter rotation

FIGURE 1.6 Black and gray connector pegs are the backbone of your Mindstorms set.

3M Connector Pegs

More pegs! The big ones shown in Figure 1.7, called *3M connector pegs,* are a lot like the black and gray ones in Figure 1.6, but they extend three standard LEGO thicknesses. In LEGO parlance, this is referred to as *3M*—the thickness of three regular LEGO beams stacked atop each other.

> **TIP**
>
> **M Refers to Length, Too**
>
> Here, the M number refers to the *length* of an element, so a 3M connector peg is the same length as three beams stacked on top of each other.

The blue pegs have friction ridges, whereas the beige ones do not. The set includes 52 blue pegs and 6 beige ones.

Beige connector pegs are smooth

Blue connector pegs have friction ridges

FIGURE 1.7 The longer connectors are delineated by both color and presence (or lack thereof) of friction ridges.

Connector Pegs with Bushing

The next type of peg is rather interesting. You may not use them all of the time, but when you need them you'll totally be grateful to have them. They're basically two-thickness pegs, but have a cross-axle bushing on the end (see Figure 1.8). This is a special connector that accommodates Mindstorms' cross axles, so you could use one to anchor the hub of a wheel, for instance. Another great use of the part is as an easily removable peg. Say you want to keep a moving part of your robot from moving; you could temporarily block the motion with one of these pegs, leaving the big part—the bushing— sticking out so you can easily grab and remove it when you're ready. You get 10 of these in the set.

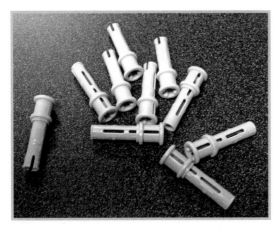

FIGURE 1.8 These connectors combine a peg with a bushing!

Connector Pegs with Cross-Axle Ends

You might have figured out by now that the Mindstorms set comes with lots of types of pegs! Figure 1.9 shows even more pegs; these pegs have a regular connector on one side and a cross connector on the other. In this book I call those *cross connectors*. You'll find yourself using these almost as much as the black ones previously shown in Figure 1.6. They're important because some LEGO beams have cross-axle holes as well as the regular round kind. As with the 3M pegs previously shown in Figure 1.7, the blue color signifies that those pegs have friction ridges, whereas the beige ones do not. The cross-axle portion is the same for both types. You'll find 24 blue pegs of this type in the set, along with 4 beige ones.

FIGURE 1.9   These pegs allow you to join two elements, one with a Technic hole and one with a cross hole.

2M Axle Peg

The 2M axle peg shown in Figure 1.10 is another commonplace one and is essentially like the ones previously shown in Figure 1.9, but these pegs are all cross axle. The 2M axle pegs are two standard LEGO thicknesses long. Nine of them come in the box.

TIP

Extricating a 2M Axle Peg

The 2M axle pegs have a tendency to get "lost" when you use them to hold two beams together. To remove one of these pegs, poke the end with a cross axle to push it out.

FIGURE 1.10   2M axle pegs.

Connector Peg with Towball

Next comes the connector peg with towball. These pegs are typically used as an end-stop for some sort of movement. If you need a LEGO part sticking out to make sure your robot's range of movement is limited, you'll want one of these pegs. Another use is as a holder for a rubber band. Say you want to have a hinged piece held taut to keep it from flapping around. Simply attach one of these pegs to each side and loop a rubber band around the towball (see the Alpha Rex's hands on the cover of your box). As you can see from Figure 1.11, the gray ones have a cross-axle connector, and the black ones have a smooth connector. You get 10 black pegs and 2 gray ones.

FIGURE 1.11   If you've ever used a ball hitch to tow a boat or trailer, these connectors will look quite familiar to you.

1/2 Connector Pegs

The final type of peg you'll find in the set is the 1/2 connector peg, shown in Figure 1.12. Just three of these pegs are included in the set. They look a lot like half of a gray connector peg. Pure and simple, these pegs provide a way of turning a beam's hole into a standard LEGO stud. You insert the long end into a hole, leaving the stud sticking out, allowing you to attach a regular LEGO brick to the stud. The connection isn't strong, and it is typically used more for cosmetic purposes than structural ones.

FIGURE 1.12   Use these pegs to transform a Technic hole into a LEGO stud.

Bushings

Next, let's look at *bushings*, which are little tubes that secure the ends of cross axles. You'll use them for pretty much every Mindstorms model you'll ever make (see Figure 1.13). Bushings come in two flavors: bushings and half bushings. The former are one LEGO thickness, and the half bushings are half as thick. With bushings, unlike most of the parts you'll find in the set, LEGO doesn't differentiate the two types with different colors, although you'll often find half bushings in bright yellow in different sets. There are 11 bushings and 9 half bushings in the Mindstorms set.

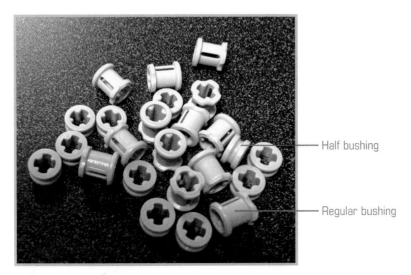

Half bushing

Regular bushing

FIGURE 1.13   Bushings are an integral part of nearly any Mindstorms model you create.

Cross Axles

Now, for the famous cross axles you've been hearing about so much. Cross axles, shown in Figure 1.14, are absolutely critical components of the Mindstorms set, used for everything from stabilizers securing multiple thicknesses of beam to, well, serving as axles for wheels. The set includes the following cross axles:

> **NOTE**
>
> **M Stands for Thickness and Length**
>
> As we discussed earlier, the "M" refers to the width of a standard LEGO brick. So, 3M is equal to the width of three standard bricks lined up side by side. But wait! Mindstorms also uses the M measurement to indicate length, so the 7M cross axle is the same length as the width of seven beams.

- 12M axles—1
- 9M axles—2
- 7M axles—4
- 6M axles—2
- 5M axles—8
- 4M axles—4
- 3M axles—19

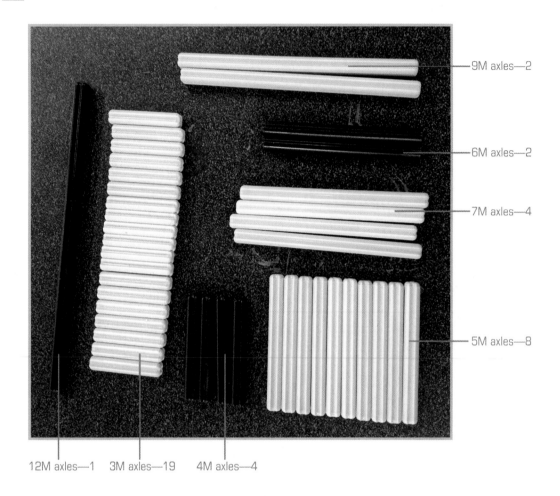

9M axles—2

6M axles—2

7M axles—4

5M axles—8

12M axles—1   3M axles—19   4M axles—4

FIGURE 1.14   Cross axles are your new best friend.

Cross Axles with End Stops

The next piece is a variant of the cross axle, which features end stops so you don't have to use a bushing to keep parts from sliding off the end (see Figure 1.15). You'd be surprised how often this piece will come in handy! The set includes three different types of axles, and each features a different type of end stop. The set contains

- 3M cross axles—6
- 5M cross axles—2
- 8M cross axles—4

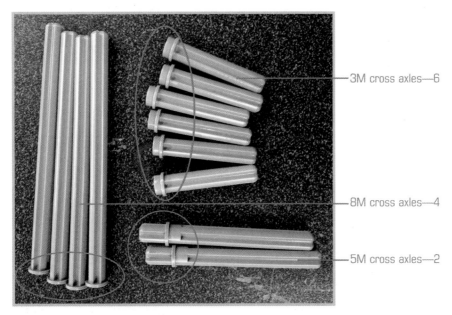

FIGURE 1.15 End stops on these axles eliminate the need for a bushing. Note that the end stops are circled in this photo.

Wheels and Treads

Wheels and treads come next (see Figure 1.16). The set contains four rims along with rubber tires. You also get two tank treads that are compatible with the rims.

FIGURE 1.16 Where the rubber meets the road.

Gears

My favorite parts of any LEGO set are the gears. I love figuring out how to mesh them together in a clever way. It is a bit disappointing, though, that LEGO includes a rather weak assortment of gears in the set (see Figure 1.17). You get a mere 11 gears:

- 36-tooth gears—2
- 20-tooth gears—2
- 12-tooth gears—2
- 12-tooth bevel gears—1
- 4-tooth gears—4

TIP

Gearheads Needn't Worry

If you, like me, yearn for more gear options, be sure to read Chapter 10, "Expanding on Mindstorms," where I talk about where to find more.

FIGURE 1.17   The Mindstorms set comes with a disappointing number of gears.

Teeth

The next parts, *Bionicle teeth,* are kind of unusual for Mindstorms in that they're almost completely cosmetic. You get 10 of them in the set (see Figure 1.18). Although they're usually decorative, another possibility might be to use them in conjunction with a motor to serve as some sort of dial, rotating to a certain direction as directed by the NXT brick. Also, in the Backscratcher Bot, our first project in this book, we use the teeth as the business end of the scratcher; they're what scratch your back!

FIGURE 1.18   These teeth are mostly for fun, but if you're clever, you can come up with other more useful applications.

Balls

LEGO includes multicolored balls with the set, used by robots either as missiles or as objects to be color-scanned and sorted (see Figure 1.19). One common type of robot to use these balls is a Great Ball Contraption. These complex linked robots are often found at LEGO conventions. Some assemblies are so huge that they cover multiple tables and spend their time rolling, pulling, lifting, and shooting the balls in an endless chain reaction. You get three of each color ball with your set.

FIGURE 1.19   These colored balls provide a variety of interesting possibilities.

Beams With Pegs

The so-called beams with pegs are essentially short Mindstorms beams with connector pegs sticking out (see Figure 1.20). You can add them as stabilizers to a wobbly model, and they're also great for changing the angle of the beams because most LEGO beams have holes on two sides only. That means you sometimes have to resort to trickery to attach parts to the smooth sides. You get six of the angle beams and 14 of the 3M beams. I wish the set included more; I use them all the time!

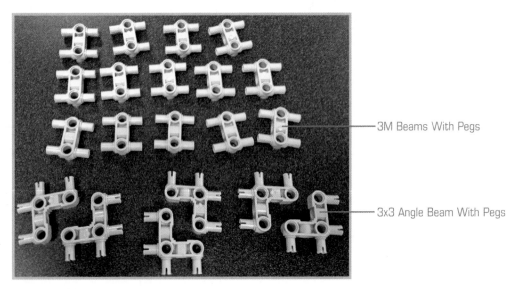

3M Beams With Pegs

3x3 Angle Beam With Pegs

FIGURE 1.20   Crafty builders make good use of beams with pegs.

Beams

Next, let's check out beams, the bones of Mindstorms robots. Using these building blocks, you can pretty much create whatever you want—with the help of the other components, of course! The kit comes with the following beams (see Figure 1.21):

- 2M beams—8
- 3M beams—10
- 5M beams—18
- 7M beams—20
- 9M beams—14
- 11M beams—6
- 13M beams—10
- 15M beams—2

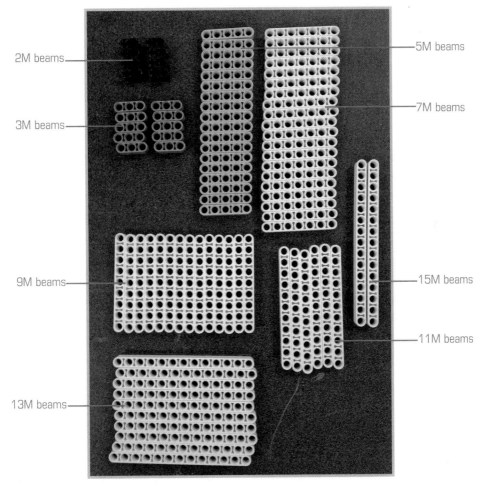

FIGURE 1.21   Beams form the skeleton of any Mindstorms bot.

Angle Beams

Angle beams add critical stability, enabling you to connect parts in ways that a straight beam cannot. For instance, you could reinforce the corners of a square robot with a 90-degree beam on each corner.

The orange parts in Figure 1.22 are the same as the similar-appearing dark gray parts, one of the few instances in the set where you get two basically identical parts sporting different colors. As you can see in Figures 1.22 and 1.23, the set includes

- 3×5 90-degree angle beams (gray)—14
- 2×4 90-degree angle beams (gray)—13
- 2×4 90-degree angle beams (orange)—4
- T-shaped angle beams—2
- 45-degree double-angle beams (medium gray)—2
- 45-degree double-angle beams (light gray)—4
- 3×3 angle beams (gray)—6
- 3×7 angle beams (white)—10
- 4×6 angle beams (white)—4

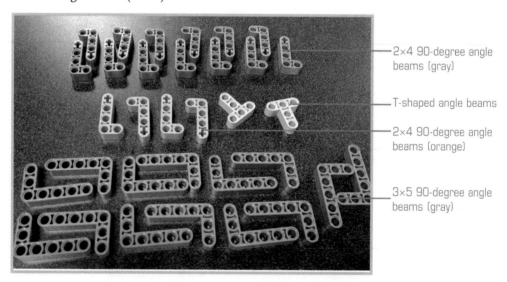

2×4 90-degree angle beams (gray)

T-shaped angle beams

2×4 90-degree angle beams (orange)

3×5 90-degree angle beams (gray)

FIGURE 1.22 L- and T-shaped angle beams provide stability and provide lots of building options.

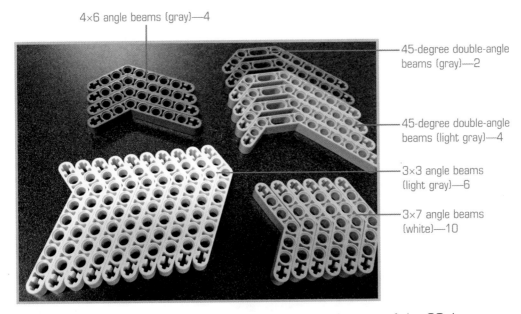

4×6 angle beams (gray)—4

45-degree double-angle beams (gray)—2

45-degree double-angle beams (light gray)—4

3×3 angle beams (light gray)—6

3×7 angle beams (white)—10

FIGURE 1.23   These angle beams help break your robot out of the 90-degree mindset.

Car Parts

The parts shown in Figure 1.24 are purely cosmetic, made to make robotic cars more cool looking but not adding a lot of value. Essentially, what you're getting are the fenders and side panels of a car's body. As with other cosmetic parts, what you get out of them depends on your own cleverness. Who knows what functional mechanisms could be created?

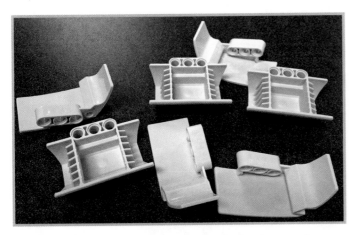

FIGURE 1.24   While purely cosmetic in nature, these parts will help you trick out any robotic cars you decide to build.

Steering Links

The parts shown in Figure 1.25 are called *steering links,* and they're used in conjunction with tow balls to form a flexible link between two elements.

FIGURE 1.25   Steering links offer a more flexible connection than beams.

Angle Elements

The parts shown in Figure 1.26, called *angle elements,* are used for connecting cross axles, cross connectors, and connector pegs. Not only can you connect two axles to make a bigger one, you can use them structurally with axles to make cubes, triangles, and so on.

- 0-degree angle elements—5
- 90-degree angle elements—12
- 180-degree angle elements—6

0-degree angle elements

90-degree angle elements

180-degree angle elements

FIGURE 1.26   Angle elements help connect cross axles at different angles.

Cross-Axle Connectors

Cross-axle connectors combine two cross axles into a longer one (see Figure 1.27). They also interface with other elements with cross ends. They're the sort of element you may not use a lot, but when you need one, you'll be grateful LEGO created them! You get four in the set.

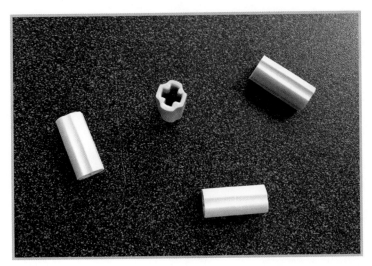

FIGURE 1.27   Cross-axle connectors make long axles out of short ones.

Cross Blocks

The parts shown in Figure 1.28 are called *cross blocks,* small beams with cross holes and Technic holes at right angles to each other. These parts allow you to attach cross axles to beams and to add perpendicular elements to help reinforce beam structures. I use the double cross blocks all the time; I don't know why LEGO includes only five in the set.

- 3M cross blocks—16
- 2M cross blocks—8
- 3M double cross blocks—5

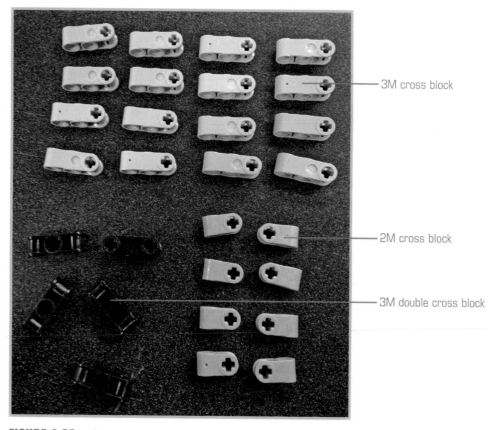

FIGURE 1.28 Cross blocks allow you to add cross-holes to beams, and at different angles.

More Miscellaneous Parts

More obscure parts! The funny-looking ones at the top of Figure 1.29 are the *magazine* and *launcher* for the ShooterBot, one of the robots you can build with the set; the instructions may be found on the Mindstorms software.

The black double pins are exactly that; you get three. The oval disks, called *cams*, are used to add an irregular motion to a motor's spin, and the four rubber things are called *rubber axle connectors*. Here's a breakdown of the oddball parts:

- Magazine—2
- Launcher—1
- 3M double Technic pin—3
- Cam—2
- Rubber axle connector—4
- Axle connector—6

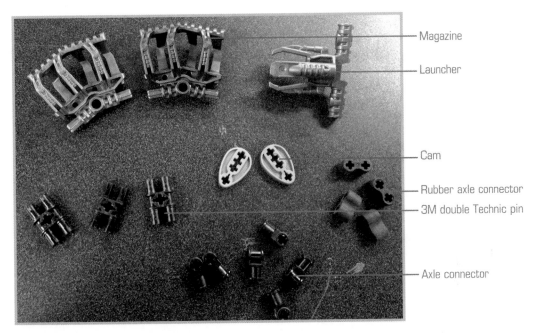

Magazine

Launcher

Cam

Rubber axle connector

3M double Technic pin

Axle connector

FIGURE 1.29 Some parts are hard to categorize. Here are some of the more obscure parts found in the set.

Peg Joiner

Peg joiners do just that, allowing you to insert connector pegs at different angles (see Figure 1.30). These parts are great for reinforcing your robot, especially if you've already built the bot and it's a little wobbly. Just reinforce the wobbly parts with some of these connectors!

- 2×3 peg joiner—6
- 2×2 peg joiner split—8
- 2×2 peg joiner—4
- 90-degree peg joiner—1

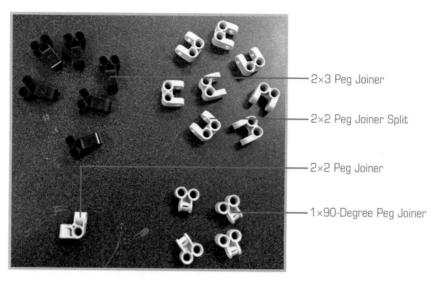

2×3 Peg Joiner

2×2 Peg Joiner Split

2×2 Peg Joiner

1×90-Degree Peg Joiner

FIGURE 1.30 Peg joiners are great for reinforcing wobbly robots.

WHAT'S WITH THAT LITTLE WHITE TILE?

It may surprise readers that LEGO included one of their classic System bricks—a white 1x2 smooth plate, shown in Figure 1.31—in the Mindstorms set. It's hard to say why it was included; it's not exactly a useful piece because it attaches via studs and none of the Technic elements in the set have studs. It's a mystery!

FIGURE 1.31 Who is to say why this one non-Technic Lego element was included?

Motors, Wires, and Sensors

The larger of the two white boxes in the main set holds all your motors and sensors, as well as the Mindstorms wires and the USB cable you use to program your NXT brick. You get three interactive servo motors, two touch sensors, a color sensor, and an ultrasonic sensor (see Figure 1.32).

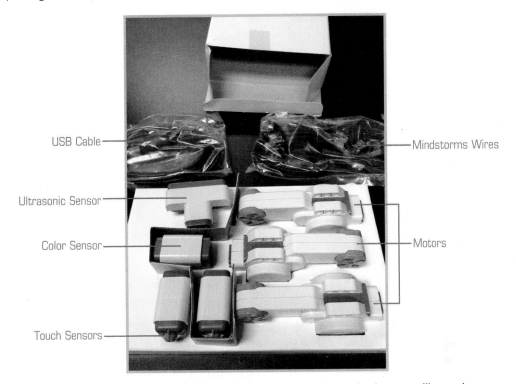

FIGURE 1.32   This box contains the sensors, motors, and wires you'll need to turn a model into a robot!

Ultrasonic Sensor

Let's take a peek at the ultrasonic sensor that comes with the set. As you can see in Figure 1.33, the ultrasonic sensor has two grills in the front, one of which covers a sensing element, while the other one emits sound. The idea is that the sensor beams out inaudible sound waves and then measures how fast the sound bounces back to the sensor, enabling you to measure distance with surprising accuracy.

FIGURE 1.33   Ultrasonic sensors emit sound waves that help your bot measure distance.

Touch Sensors

The Mindstorms touch sensors double as both pushbuttons as well as sensor-detecting contacts (see Figure 1.34). The sensors can discern three different actions: a quick bump, a press-and-hold, and a release action. You get two touch sensors with the set.

FIGURE 1.34   Touch sensors can serve as a pushbutton or as a contact sensor.

Interactive Servo Motors

Three of LEGO's interactive servo motors come with the set, and you'll probably wish there were more (see Figure 1.35). What differentiates them from motors one might find in a hobby store is that they are equipped with position encoders that enable the NXT brick to accurately determine the speed and precise angle of the motor's turn.

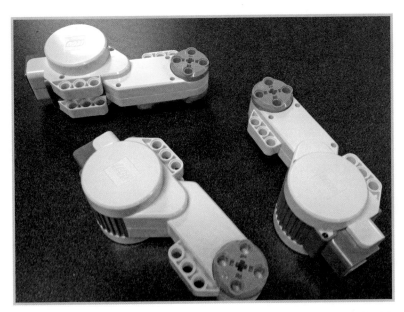

FIGURE 1.35 Three interactive servo motors are included with the NXT 2.0 set.

Color Sensor

The color sensor shown in Figure 1.36 can distinguish between colors—hence the name—but it does so much more. For instance, it not only can determine the color of a scanned object by bouncing a light off it and measuring the returned values but also can detect the level of lighting in a room (dark, poorly lit, bright, and so on). If that weren't enough, you can turn it into a lamp shining red, blue, and green lights. The color sensor is a vast improvement over the previous Mindstorms set, which only boasted a light sensor.

FIGURE 1.36   The color sensor is a multipurpose part that you'll find many ways to use.

USB Cable

LEGO includes a standard USB cable with the set (see Figure 1.37). It's fairly universal, so if you lose yours, you can always find another without buying an official LEGO product. You'll need this cable when you start programming your NXT brick.

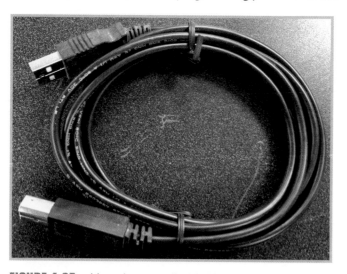

FIGURE 1.37   Use the supplied USB cable when you start programming your NXT brick.

Mindstorms Wires

The Mindstorms sensors and motors communicate with the NXT brick with the help of six-ply wires (see Figure 1.38). The set comes with the following selection of wires, giving you a nice variety for all your projects:

- 20cm wire—1
- 35cm wires—4
- 50cm wires—2

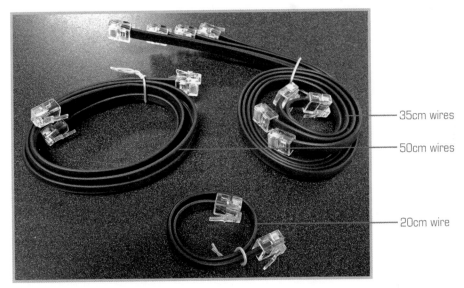

——— 35cm wires

——— 50cm wires

——— 20cm wire

FIGURE 1.38   These wires carry the communications from the NXT brick to your bot's motors and sensors.

The NXT Brick

The *NXT brick* is the brain of the robot, a computer that interprets data from the sensors and sends power to the motors to make them turn (see Figure 1.39). You learn all about the NXT brick in Chapter 3, "Anatomy of the NXT Brick," so be sure to check it out!

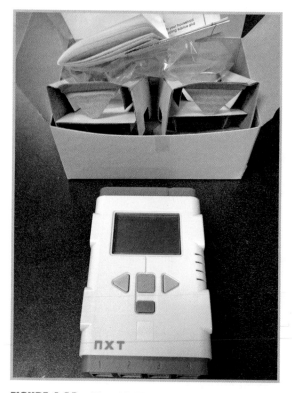

FIGURE 1.39   The NXT brick—the brain of your creations.

Next Chapter

In Chapter 2, "Project: Backscratcher Bot," we'll build our first bot—the backscratcher bot!

Project: Backscratcher Bot

Let's face it, the LEGO Mindstorms NXT set is really big and kind of intimidating if you're just getting started. Getting from opening the box to having a functional robot can be a leap, and I bet some users never even make it that far.

With the Backscratcher Bot shown in Figure 2.1, however, we're simplifying the build down to the point where it'll be a breeze for anyone to tackle this project. Not only is this a simple robot, but we'll skip the programming step—for now—so you can have a working robot within minutes. How is this accomplished? We'll use the NXT brick's menu system to control the motor without uploading a program first.

But what does the robot actually do? Basically, it's a motor that spins at the touch of a button, and the motor's hub sports a wheel studded with teeth that soothe even the most persistent of itches. More importantly, it's a quick and easy entry into the world of Mindstorms!

FIGURE 2.1 The Backscratcher Bot will fulfill your need for an automated backscratcher.

Let's do it!

Adding Batteries to the NXT Brick

Before we begin, however, let's put batteries into the NXT brick. We're going to do this for the very best reason: because it won't work without them!

First, remove the NXT brick (shown in Figure 2.2) from the Mindstorms box. It's in a smaller, white cardboard box. (If you're really curious about the rest of the stuff, be sure to read Chapter 1, "Unboxing the LEGO Mindstorms NXT Set," where we unbox the entire set!)

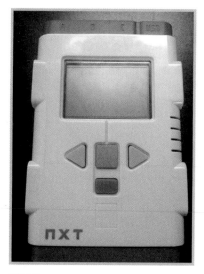

FIGURE 2.2   The NXT brick as it appears out of the box.

Next, squeeze the tab on the back of the NXT brick to release the battery cover (see Figure 2.3). It should pop off easily; if it gets stuck, squeeze the tabs a little more firmly.

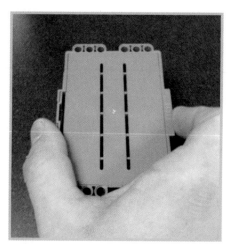

FIGURE 2.3   Flip over the brick to get at the battery cover.

When the brick is open, you can see the carrier for the batteries (see Figure 2.4). You need eight standard AA batteries. If you use rechargeable batteries, be aware that they output less juice than regular batteries. If you want to get technical, rechargeables use 1.2 volts versus the 1.5 used by regular batteries. That means your motors will move slower than you might expect.

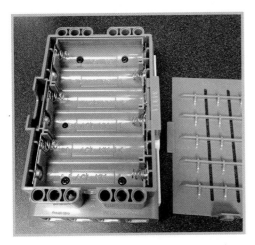

FIGURE 2.4   So that's where we keep those batteries!

Place the batteries as you see in Figure 2.5, with each cell flipped opposite its neighbor. The flat end of each battery, marked with a minus sign, goes up against the spring.

FIGURE 2.5   Pop in the batteries as you see them here.

Reattach the cover and you're done! When the battery cover is replaced, the unit automatically powers up and emits a startup tune (see Figure 2.6).

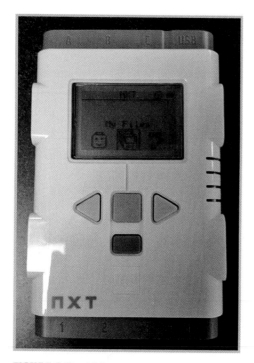

FIGURE 2.6 When the unit is powered up, the menu appears.

Parts You Need

Next, let's build the actual robot. Dig through the Mindstorms set and find the parts you see in Figure 2.7. Table 2.1 shows a parts list and the number of each you should have. Again, if you're curious about the stuff you find in the parts bags, check out Chapter 1, where I run through everything included in the set.

TABLE 2.1 Parts List for the Backscratcher Bot

Quantity	Part
1	1NXT Brick
1	Interactive Servo Motor
20 cm	Wire (Not shown in Figure 2.7)
8	Teeth
2	36-Tooth Gears
1	5M Beam
3	7M Beams
2	15M Beams

2	Double Angle Beams 3×7
2	Angle Beams 3×5
24	Connector Pegs (black)
2	3M Connector Pegs (blue)
8	Cross Connector Pegs (blue)
4	Bushings
2	Double Cross Blocks (black)
2	Cross Blocks (gray)
4	3M Cross Axles
2	5M Cross Axles

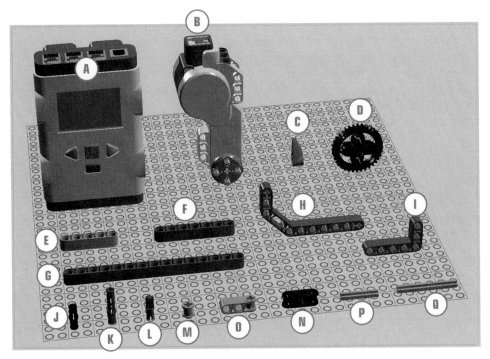

FIGURE 2.7 Bricks you need to build the Backscratcher.

A. 1 NXT brick
B. 1 interactive servo motor
C. 8 teeth (orange)
D. 2 36-tooth gears
E. 1 5M beam
F. 3 7M beams

G. 2 15M beams
H. 2 double angled beams 3×x7
I. 2 angle beams 3×x5
J. 24 connector pegs (black)
K. 2 3M connector pegs (blue)
L. 8 cross connector pegs (blue)

M. 4 bushings
N. 2 double cross blocks (black)
O. 2 cross blocks (gray)
P. 4 3M cross axles
Q. 2 5M cross axles

Step-by-Step Assembly Instructions

Let's begin. Grab the parts you pulled from the bags and follow these steps. Assembling the bot is super easy!

  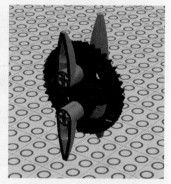

The parts added in this step are shown in blue

Orange is the actual color of the part

STEP 1 Put two 3M cross axles into the 36-tooth gear. They should stick out equally on both sides. To make sure you grabbed the right cross axle, note that the length of the rods are exactly three times the width of the gear—thus, the 3M!

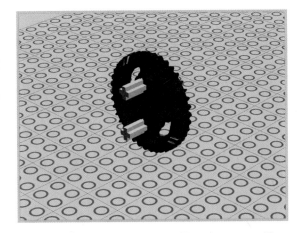

STEP 2 Add the teeth, two on each side.

STEP 3 Make another assembly just like it. That's right; you'll need two gears, each sporting four teeth.

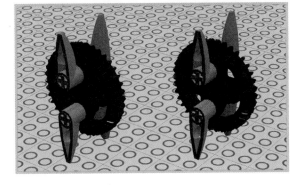

STEP 4 Insert two 5M cross axles into the motor's orange hub and add a bushing to each side. The bushings serve as spacers. Arrange them exactly as you see here.

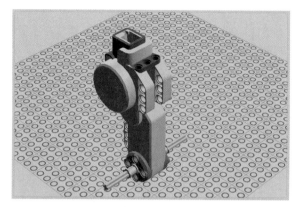

STEP 5 Add the two gear assemblies you created in steps 1–3.

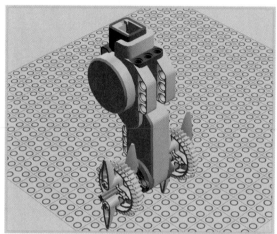

STEP 6 Secure the gears with a pair of bushings.

STEP 7 Grab the cross blocks and arrange them as you see here. Add connector pegs to the cross blocks.

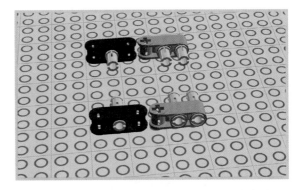

STEP 8 Add blue connector pegs, the ones with a cross-shaped profile on one side, to the remaining holes on the cross blocks.

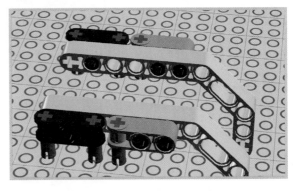

STEP 9 Connect the two angle beams to the black connector pegs sticking out of the cross blocks.

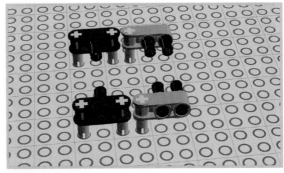

STEP 10 Add black connector pegs to the angle beams.

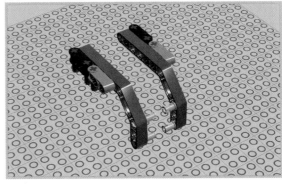

STEP 11 Connect the liftarms to the motor assembly you already built.

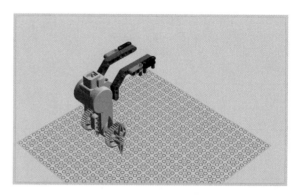

STEP 12 Add your 15M beams to the blue pegs.

STEP 13 Attach four black connector pegs as you see here.

STEP 14 Connect two 7M beams to the black pegs.

STEP 15 Connect another 7M beam with the help of two 3M blue connector pegs. Shove them through the holes from above; then attach the beam below.

STEP 16 Add six black connector pegs like you see here.

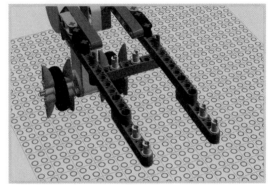

STEP 17 Place the 5M beam and the angled beams onto the connector pegs you just attached.

STEP 18 Yay, six more connectors!

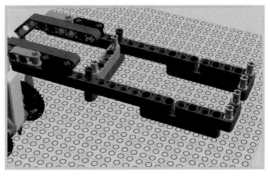

STEP 19 Connect the NXT brick. Notice at the base of the brick how its curves match the angle beams.

STEP 20 Add the wire and you're done! Connect it from the motor to the B port on your NXT brick. (You can find the wire in the same box as the motor.) Note that our design program doesn't show wires; the green arrows indicate that the wire continues between those points.

WHAT THE HECK IS THIS?

When you dug through your Mindstorms box, you likely encountered this bag. It contains the parts you need to build a simple rolling robot, shown on the cover of the user guide. It's sort of a basic model that can you build into a more advanced robot—the ShooterBot—which is one of the four models that LEGO has you build with the Mindstorms set. You can build it if you want; the instructions are in the *Mindstorms User Guide* that came with the set.

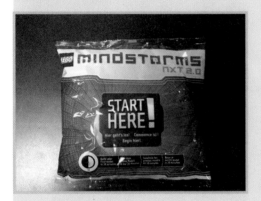 

Programming the Backscratcher Bot

Next, let's use the NXT brick's built-in programmer to control the motor and therefore make the backscratcher actually scratch backs. Is this Mindstorms programming? Not really. There is a much more robust programming environment available on the software disc you got with your Mindstorms set. However, this simple system included on the NXT brick's firmware lets you quickly whip up a simple program. Let's do it!

STEP 1 If it's not on already, you need to power on your NXT brick by pressing the orange button on the front of the brick. The LEGO logo shows on the screen and a fun little song plays. The NXT brick offers a simple programming interface that requires no computer to work.

STEP 2 Press the right-arrow button to get to the NXT Program menu; then press the orange button to confirm your selection.

STEP 3 When you get to the intro screen, press the orange button again.

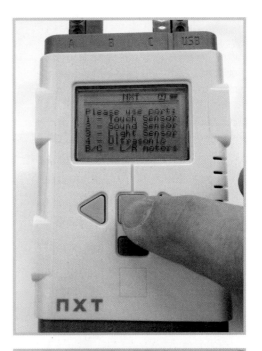

STEP 4 Now you get to the main programming screen. As you can see, you have five steps at the top of the screen and some options for what to put into those steps at the bottom of the screen.

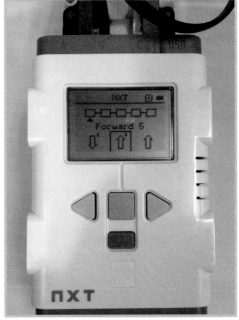

STEP 5 The first step defaults to "Forward 5." Press the right-arrow button to change the first step to "Forward" and then press the orange button to lock it in. The programmer automatically advances to the second block.

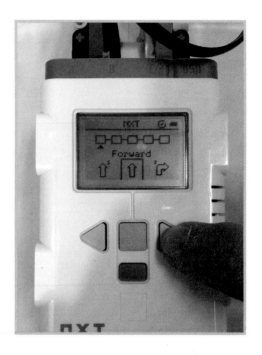

STEP 6 It defaults to "Empty," so leave it that way. Press the orange button to continue. Not sure what step you're on? Keep an eye on the little arrow; in this case, it's underneath the second block.

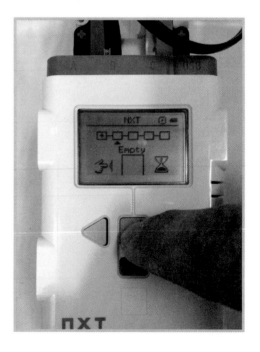

STEP 7   Here, the program defaults to "Forward 5." Press the right-arrow button to select "Forward" and then lock it in by pressing the orange button.

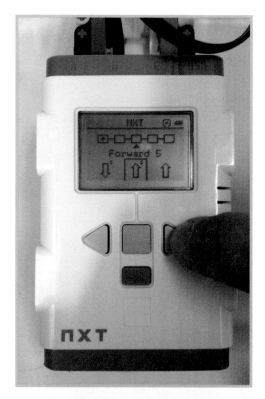

STEP 8   The program now defaults to Empty. Keep it that way by pressing the orange button.

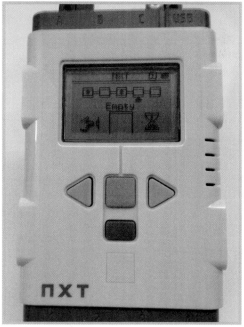

STEP 9 You have the option to loop the program, meaning having it continuously run. It defaults to "Stop," but you don't want the program to end when it reaches the fifth step, so press the right-arrow button to change it to "Loop." Then lock it in by pressing the orange button.

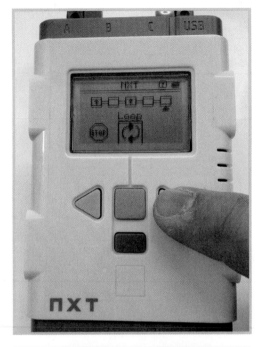

STEP 10 Finally, you're ready to run or save the program, or alternatively to abandon it and return to the main menu. If you select "Save," you have the option to save the file to your NXT Programs menu. If you choose to run the program without saving, the program will run as long as the brick is powered on. However, after you power off, the program is lost, so remember to save it if you want to keep it! If you accidentally forget to save it, you have to program it again.

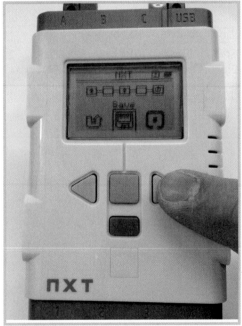

STEP 11 Next, let's name the program. The filename defaults to "untitled," but we can do better than that! Use the arrows to toggle between the various letters and numbers, using the orange button to select each letter. If you press the wrong one by accident, delete it with the gray button. When you're done, toggle back to the check mark to confirm your filename.

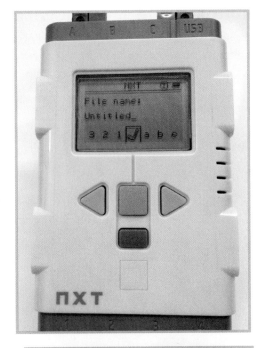

STEP 12 You're done! The NXT Menu should look like this.

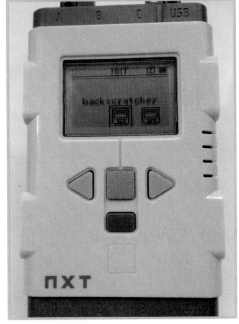

You're done! You're now in possession of a motorized backscratcher you made yourself. Go forth and scratch backs, and think about what you learned in this chapter.

Running the Backscratcher Bot

OK, now what? First, find a back in need of scratching. Next, power up your NXT brick and toggle the left button once to get to NXT Files. Select that and find your backscratcher program, and then select it. You're in business (see Figure 2.8)!

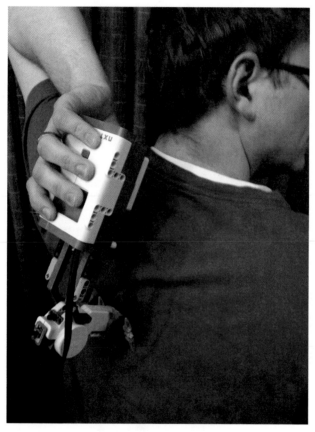

FIGURE 2.8   Ahhh, much better.

Next Chapter

In Chapter 3, "Anatomy of the NXT Brick," we'll delve into the NXT brick, examining every menu option in detail so you can learn how to use this great tool.

Anatomy of the NXT Brick

It's difficult to exaggerate how important the NXT brick is to the Mindstorms set. Heck, LEGO even put "NXT" in the name of the product (see Figure 3.1)! The brick controls all the sensors and motors of your robot and can even communicate with other NXT bricks. Essentially, the NXT is a mini-computer very similar to a PC, only less powerful.

In this chapter, we examine the brick itself, as well as how to interact with it via its built-in screen and control buttons. Want to learn how to actually program it? You have to wait for Chapter 4, "Introduction to Programming"!

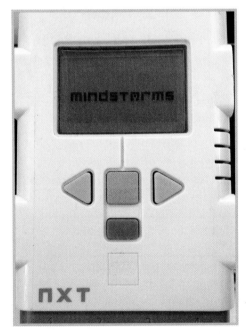

FIGURE 3.1 The NXT intelligent brick is the most complex and fascinating element of the Mindstorms set.

The Brick

First, let's examine the NXT brick's physical features (see Figure 3.2). After that, we go through its menus.

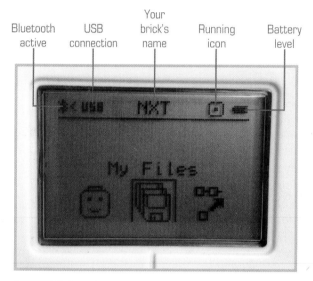

| Bluetooth active | USB connection | Your brick's name | Running icon | Battery level |

FIGURE 3.2 The icons at the top of the NXT's screen tell you what's going on with the brick.

The NXT brick's screen—and the accompanying buttons—is the primary way to manage the software files that come with the NXT, as well as those that you upload to the brick itself. The screen is a black-and-white, 100×64-pixel LCD with no backlighting.

The screen features some cryptic icons (see Figure 3.2). Refer to the following key:

- **Bluetooth**—When the Bluetooth (wireless connectivity) feature of the brick is activated, the icon appears. See the "Bluetooth" section later in this chapter for more information about this technology.
- **USB**—This icon shows that you've successfully connected the NXT brick to your computer via the USB cable. If you have it plugged in but the icon hasn't appeared yet, check the connection!
- **NXT name**—Give your brick a unique name! This is mostly for the cool factor but can be helpful if you have multiple NXT bricks and are trying to keep track of which one is which, especially when connecting wirelessly. You must change the name through the software that came with the set; you can find more information on this in Chapter 4. The name defaults to "NXT."
- **Running icon**—When your NXT brick runs, this icon spins. If it ever stops spinning, your NXT has crashed! See the "Reset Button" section for a guide on how to fix this glitch.
- **Battery**—This icon lets you know the battery level of the brick. When you need to replace the batteries, the icon flashes.

Buttons

You access the NXT brick's menus through four buttons (see Figure 3.3). They allow you to advance through the options and select the one you want or go back to a previous option. Even better, you can use the buttons to control a program while it runs; see the "NXT Buttons Block" section in Chapter 4 to learn how to do that. The buttons consist of the following:

FIGURE 3.3   A surprising amount of functionality is built into these four rather innocuous-looking buttons.

- **Orange button**—This button turns on the NXT brick if it's shut off and also is the "Enter" button for the NXT brick. When you find an item in the menu you want to select or run (such as a program), you press the orange button to trigger that action.
- **Arrows**—You can navigate through the various menus and options using the left and right arrows.
- **Gray button**—This button is the "go back" button if you make a mistake or select something you didn't mean to. You can also shut down the NXT brick's power from the main menu by pressing this button (more than once if you're deep into a set of onscreen commands) and then the orange button to confirm.

Ports

The NXT communicates with its sensors and motors via wires plugged into the NXT brick's ports. A glance at the top of the brick (see Figure 3.4) shows three motor ports labeled A, B, and C on the top of the casing, along with a USB port conveniently labeled "USB."

NOTE

Motors Plug Into Letters; Sensors Plug Into Numbers

The lettered ports accept *only* motors. If you try to plug a sensor in to A, B, or C, it doesn't work. Similarly, you can only plug sensors into ports 1, 2, 3, or 4, but they won't work in the lettered ports.

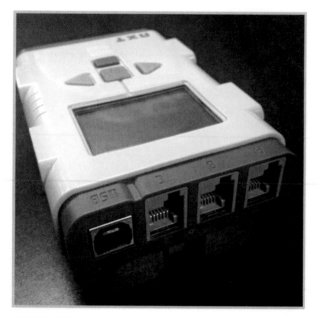

FIGURE 3.4   The NXT's top features the motor ports and USB input.

The bottom of the brick (see Figure 3.5) features four ports labeled 1, 2, 3, and 4. These four accept only sensors and can't power a motor.

FIGURE 3.5   The NXT's bottom sports the sensor ports.

Connector Holes

The NXT brick, like all proper LEGO elements, enables you to incorporate it into your robot with the help of a variety of holes that accommodate the Mindstorms connector pegs. There are 12 connector holes on the bottom and 10 on each side, and (as usual) they completely match the holes found on Mindstorms beams. Figures 3.6 and 3.7 show these connector holes.

Connector holes along the right side of the NXT brick

FIGURE 3.6   The right side of the NXT sports 10 connector holes.

Connector holes on the bottom side of the NXT brick

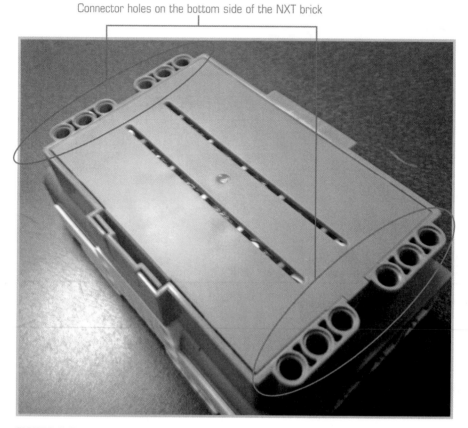

FIGURE 3.7   The bottom also has a bunch of connector holes.

Reset Button

Flip the brick over and look inside one of the connector pegs along the top edge, closest to the USB port. There, you'll see a tiny button for resetting the NXT if it crashes (see Figure 3.8). If your NXT brick freezes up and won't respond to button presses, doesn't make the sounds you're expecting, or if the running icon stops spinning, just press the reset button with a toothpick, paper clip, or other narrow pointy thing. Doing so will automatically power down the brick and restart it.

NOTE

Using the Reset Button

If your brick stops responding, press the reset button with a paperclip and the NXT will automatically restart with all of your programs and files intact.

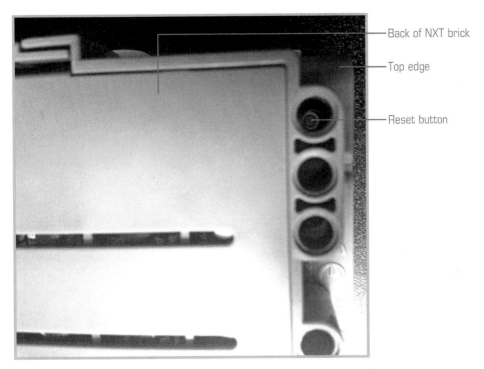

Back of NXT brick

Top edge

Reset button

FIGURE 3.8 Look inside the upper hole. What do you see?

Menus

The heart of the brick is its software, accessible through a series of menus. You can navigate these menus by using the arrow buttons to page through each screen. When you find an option you want, just press the orange button to select it. If you accidentally select the wrong menu item, press the gray button to back out. The following sections describe the various options you'll find.

My Files

The My Files menu organizes all your programs for easy access (see Figure 3.9). You could, therefore, hold multiple variants of a robot's program in the brick and navigate between them to test various subsystems. The NXT divides your programs into various categories: Software, NXT Files, Sound Files, and Datalog Files.

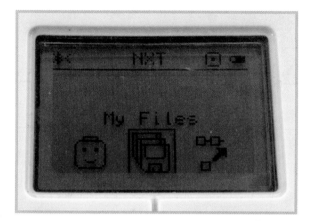

FIGURE 3.9 The My Files menu organizes the programs, sound files, and data stored on the brick.

Software Files

Software files are programs that you have downloaded from your computer to control your robot. Say you create a program in NXT-G, the Mindstorms default programming environment. When you send it to the NXT, it appears in the Software Files menu (see Figure 3.10). Simply use the arrow buttons to scroll through the icons until you find the one you want.

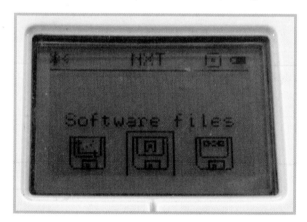

FIGURE 3.10 The Software Files menu is the place where you can find the programs you have downloaded from your computer to your NXT brick.

NXT Files

NXT files are programs created on the brick itself with the programming interface that you used in Chapter 2, "Project: Backscratcher Bot," to program the Backscratcher. When you complete such a program, simply save it and it will appear in the NXT Files menu (see Figure 3.11).

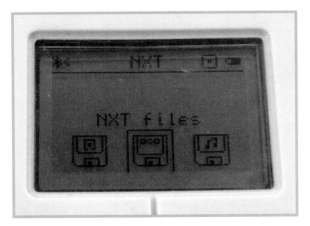

FIGURE 3.11 The NXT Files menu shows the programs that you created directly on the NXT brick, as opposed to downloading them from your computer to the brick.

Sound Files

You can upload sound files to enhance your robot. For instance, you could program your brick to trigger an alarm after it completes a maze. The NXT comes with several sound files already, some of which are the system sounds (the startup song, for instance). Figure 3.12 shows the Sound Files menu.

CAUTION

Delete with Care

If you delete one of these sound files, the NXT still functions but doesn't emit the sound you're expecting. To get those files back, you'll have to reload the brick's firmware—the default programs LEGO loads onto the NXT. I'll show you how to do this later on in the chapter.

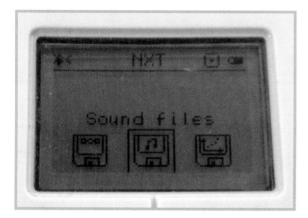

FIGURE 3.12 Associate sounds from this menu with specific actions.

Datalog Files

Finally, there are datalog files. Some robots—mainly those with sensors—can create records of the levels the sensors return. For instance, a robot could track the distance between itself and the nearest wall, or the amount of light the light sensor is detecting. Records generated by the datalog function are stored here (see Figure 3.13).

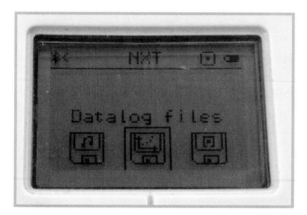

FIGURE 3.13 Data collected by your robot is stored here.

NXT Program

Figure 3.14 shows the NXT Program menu, which allows you to access the various programs (if any) you created using the NXT's on-board programming tool. To recap, the NXT allows you to make very simple programs just using the NXT's buttons, allowing you to play with a robot without tweaking a program, or to test a motor you think might be burnt out. As we explored while programming the Backscratcher Robot in Chapter 2, you basically have five "steps" that can each be set to perform a certain action such as turning a motor's hub forward or backward. Once you've finished the program it can be run as-is or saved to the NXT Files menu and launched from there.

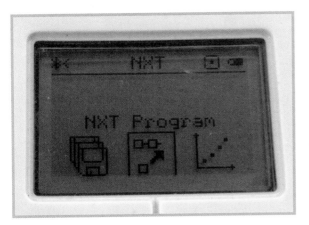

FIGURE 3.14   The NXT Program menu is the place where you program your bot to do thy bidding

Try Me

The Try Me menu allows you to test motors and sensors in a fun way (see Figure 3.15). Here's how it works. First, plug a motor or sensor into a port. Don't worry; the menu tells you the correct port if you get it wrong. Then press the orange button to hear a fun sound as the NXT turns the input from the sensors or a turn of the motor's hub into noises coming from the speaker. Try Me is fairly useless but might come in handy if you want to, for instance, test a sensor you suspect might be nonfunctional.

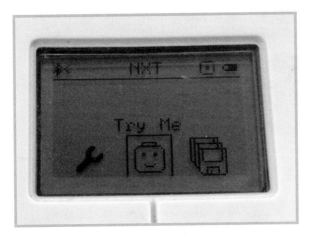

FIGURE 3.15   The Try Me function can help you troubleshoot motors and sensors.

The View Menu

Like the Try Me software, the View menu allows you to interact with the sensors and motors. However, View is different and cooler than the Try Me function (see Figure 3.16).

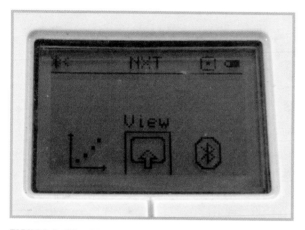

FIGURE 3.16 The View menu provides some useful diagnostic capabilities.

Rather than merely generating sounds with them, the NXT brick actually pulls in sensor data and gives you a numerical measurement or reading. Here's how:

1. Plug a motor or sensor into any of the appropriate ports (that is, only a motor in ports A–C and only a sensor in ports 1–4).

2. Select the type of sensor you'd like to use from the View menu (see Figure 3.17).

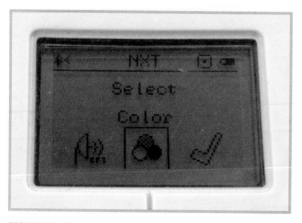

FIGURE 3.17 Select your sensor type. In this case, we chose the color sensor.

3. Choose the appropriate port from the list the menu supplies you (see Figure 3.18).

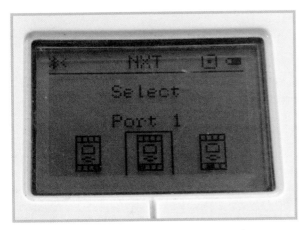

FIGURE 3.18   Select whichever port your sensor or motor is plugged into.

4. Choose what data you want to collect, such as
   - The reading from the sensor
   - What color is detected by the color sensor
   - What measurement the ultrasonic senor uses (inches or centimeters)
   - Number of rotations of the motor

Figure 3.19 shows the result of our color scan: black! The View menu is such a great way to play around with the sensors.

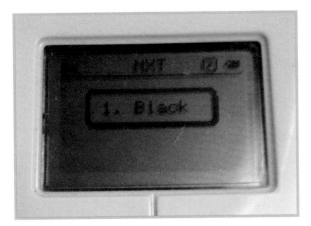

FIGURE 3.19   Your reading appears on the screen!

NXT Datalog

The NXT Datalog program stores data from your sensors (see Figure 3.20). For example, you could have the brick keep track of the readings from the light sensor and store that data in the Datalog Files section. Now, why would you want to do such a thing? Imagine the purpose of your robot is to perform an experiment—for instance, recording the light levels of a room—and rather than having a human standing there with a clipboard, you can set the robot to record its own readings. The resulting data can be uploaded to a PC for more detailed analysis.

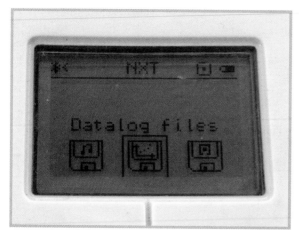

FIGURE 3.20   Data from your bot's sensors is stored here.

Settings

The Settings menu has some useful features for managing your NXT brick (see Figure 3.21). Here's what you can do:

CAUTION

Mayhem Could Ensue

Be careful! One of the items on the Settings menu enables you to delete all your files. If you do this by accident, all default system files can be restored by reloading the NXT's firmware from your NXT software. I'll explain how later in this chapter.

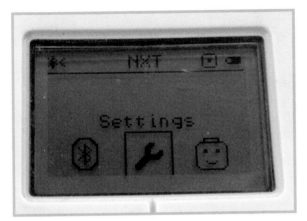

FIGURE 3.21 The Settings menu provides controls for managing your NXT brick.

Volume

The Volume menu controls the loudness of the noises the NXT brick makes, including both the system noises (the startup song and the button click) as well as any noises your robot makes (see Figure 3.22). The volume setting ranges from 0 (mute) to 4 (maximum volume). Conveniently, when you turn off the NXT brick, it doesn't forget your favorite setting.

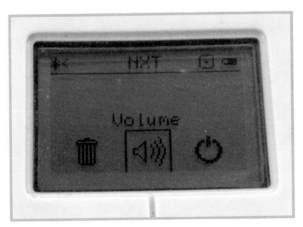

FIGURE 3.22 The volume setting controls how much racket your bot makes.

Sleep

The Sleep option sets how long the NXT brick sits idle before it automatically powers off (see Figure 3.23). You can choose one of six settings between 0 (never shuts off) to 60 minutes. Setting it to never shut off is really not a good idea. If you forget to manually shut down, you could be faced with eight dead batteries the next time you try to power on the NXT.

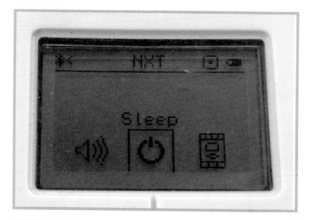

FIGURE 3.23   Your bot conserves its power by going to sleep if left idle.

NXT Version

The NXT Version item displays the firmware version and various other obscure information about the NXT (see Figure 3.24). The only way I could see you using this is if you don't update the NXT's firmware for a while and are curious about whether it has the latest version. I explain how to update the NXT's firmware later this chapter.

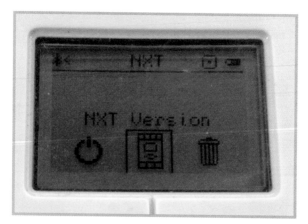

FIGURE 3.24   Use this item to see what version of the NXT firmware your brick is running.

Delete Files

You may find it shockingly easy to delete seemingly important files off the NXT brick. The Delete Files option under the Settings menu (see Figure 3.25) essentially allows you to delete any of the data files from your NXT—even all of them at once! Even files the system uses such as the sound files will vanish if you're not careful, or you can lose all the data the data-logging feature has gathered.

TIP

Replacing Deleted System Files

You can replace deleted system files by refreshing the NXT firmware through the software interface. I show you how to do that later this chapter.

FIGURE 3.25 Tired of all those files cluttering up your NXT? Just delete them.

Bluetooth

Bluetooth allows you to wirelessly transmit data; the same technology is used in wireless keyboards and cell phone earpieces. It turns out that the NXT brick has a Bluetooth chip in it, allowing you to connect the brick to your computer or to other NXT bricks. The Bluetooth menu option turns on Bluetooth and controls the connection process with other devices (see Figure 3.26).

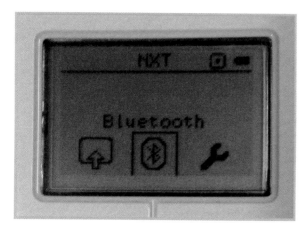

FIGURE 3.26   Bluetooth allows your bot to wirelessly transmit data.

In addition to being able to send files from your computer to the NXT and from the NXT to another NXT, you can even control your robot wirelessly as if your PC were a remote control! Here's how you configure your Bluetooth settings:

1. Turn on Bluetooth (see Figure 3.27). You should leave it off if you're not planning on using it, because it's a battery drain.

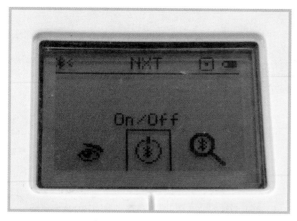

FIGURE 3.27   Power on Bluetooth with this option.

2. Search for another Bluetooth-enabled NXT brick or computer (see Figure 3.28).

FIGURE 3.28 What Bluetooth devices are nearby? Search and find out.

3. If you've previously paired with another brick or computer, it'll show up in your Contacts list (see Figure 3.29).

FIGURE 3.29 The Contacts menu option lists previous connections.

4. The Connections menu item (see Figure 3.30) shows you the number of active connections, enabling you to potentially connect to more than one Bluetooth device.

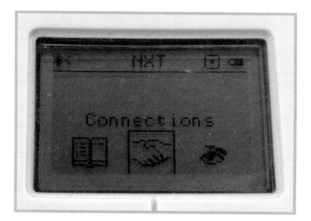

FIGURE 3.30 See what Bluetooth devices are paired with the NXT by looking at the Connections feature.

5. Visibility (see Figure 3.31) is a privacy setting enabling you to have Bluetooth on without your NXT appearing on other people's lists of accessible devices.

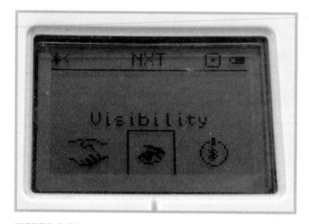

FIGURE 3.31 Decide whether your NXT will be visible to other Bluetooth-enabled devices.

ESTABLISHING A BLUETOOTH CONNECTION

Here's how you can connect two Bluetooth devices:

1. Make sure both units are powered on, and that Bluetooth is activated on both.

2. In the Bluetooth menu, select Search. It may take a few moments for the NXT to iden-tify other Bluetooth-enabled devices in the area.

3. When the search has completed, select the second NXT or the computer from the list. (This is where it would be helpful to name your NXT brick, because this name is what shows up to identify the brick. Having six bricks—hypothetically—named "NXT" might get a little confusing!)

4. After the two devices make contact, you'll need to come up with a passkey to ensure that you're connecting to the right device and no unauthorized devices can connect. Enter the passkey in both devices and you're golden!

For more detailed information on how NXT works with Bluetooth technology, read pages 36–45 of the user guide or check out www.mindstorms.com/bluetooth.

Powering Your NXT

The NXT runs off batteries, so monitoring its power situation is important. Here are some tips for conserving power:

- **Shutoff**—You can shut off your NXT brick at any time by navigating back to the top level of the menu system by pressing the gray button repeatedly, and then the orange button to confirm when it asks you if you want to shut down (see Figure 3.32).

- **Sleep**—As mentioned in the "Settings" section of this chapter, you can set how long the NXT will sit idle before it shuts itself off.

- **Using rechargeable batteries**—You may get tired of buying lots of AA batteries and want to find some rechargeables. However, don't go for regular rechargeable AAs. Instead, check out the optional battery pack (product number 9693) available from the LEGO store. It fits into the NXT brick in place of the AAs.

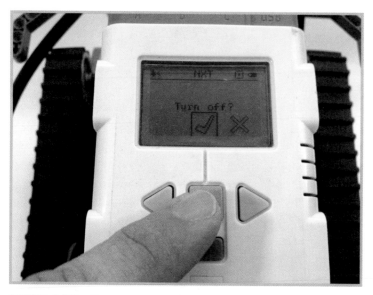

FIGURE 3.32 To shut off your NXT, press the gray button repeatedly until this screen appears, then hit the orange button.

Resetting a Crash

If your NXT brick crashes, reset by pressing the button hidden in the hole shown in Figure 3.8 earlier in this chapter. You'll probably have to use something slender like a paper clip to press the button. Triggering it restarts the brick.

CAUTION

Don't Hold It Too Long!

When resetting your brick, make sure you don't hold down the button for more than four seconds; otherwise, it will wipe the NXT and you'll have to update the firmware.

Updating NXT Firmware

As mentioned earlier, the firmware is the software that controls the NXT brick. Sooner or later, you'll delete something important off your NXT, or simply will realize that you're a version behind and need to update the firmware. Here's how you do it:

1. First, make sure your NXT brick is powered on and plugged into your computer with a USB cable—you can't update over Bluetooth.

2. Launch the Mindstorms software and select the 'Update NXT Firmware' option from the Tools menu (see Figure 3.33).

3. Click the Online Updates option. This will take you to LEGO's website where you can see what updates are available. If you spot a version number (hint: They start with a V, as in V1.31) higher than one you have, download it.

4. You'll need to manually put the downloaded firmware in the right folder.
    - On PCs, it should go here: C:\Program Files\LEGO Software\LEGO Mindstorms Edu NXT\engine\Firmware.
    - If you're on a Mac, put it here: Applications > LEGO Mindstorms NXT > engine > Firmware.

5. The latest versions will appear under Available Firmware Files. Select the latest firmware (the one with the highest number) from the list. (Figure 3.33 just shows one file, v1.28.)

6. Select the file you want and press download. When the three progress bars are finished, the firmware is upgraded!

FIGURE 3.33 Update your NXT easily by using Mindstorms' firmware tool.

The Next Chapter

In Chapter 4, "Introduction to Programming," we keep delving into the NXT and learn how to actually program it. This will give us the foundation to tackle ever more challenging projects!

Introduction to Programming

In Chapter 1, "Project: Backscratcher Bot," you learned how to use the NXT brick's on-board programming tool to make the Backscratcher Bot's motor turn. Now we're going to learn how to program the robot properly, using NXT-G, the software that comes with the Mindstorms set. Figure 4.1 shows the NXT-G welcome screen.

NOTE

This Is a Primer

This chapter is a primer that introduces you to the basic Mindstorms programming environment. We don't get into much in the way of hands-on programming here, although we do get into the specifics within each of the project chapters and in Chapter 8, "Advanced Programming."

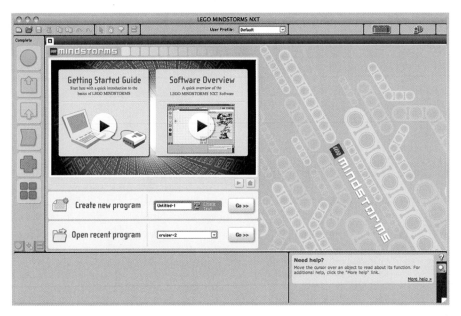

FIGURE 4.1   The welcome screen of the Mindstorms software.

System Requirements

NXT-G isn't a particularly robust program, meaning it can run on older computers that probably ought to have been recycled long ago. Nevertheless, it does have minimum requirements that must be met for successful installation.

- **Linux**—Let's get this out of the way: NXT-G doesn't run on Linux unless you have an emulator such as WINE (http://www.winehq.org/). If you don't have this (or a similar) program or don't want it, you have to use another machine to run NXT-G—one with either Mac OSX or Windows.
- **Windows**—NXT-G expects a minimum of Windows XP Professional or Home Edition with Service Pack 2, or a newer version of Windows such as Vista, Windows 7, or Windows 8.

  For XP, you need a minimum of an 800MHz processor, although 1.5GHz is recommended. Vista and higher users should have Service Pack 1 installed and a 1GHz processor or better.

  Additionally, you need a CD drive, a minimum of 512MB RAM, 700MB available hard disk space, a USB port, and a 1024×768 display.
- **Macintosh**—Your Mac should have OS 10.4 or newer, and run either a PowerPC processor (G3 and 600MHz or better) or an Intel processor (1.3GHz or better). You also need a DVD drive, 512MB RAM, 700MB of drive space, a USB port, and a 1024×768 display.

Installing the Software

Find the CD that came with the set (see Figure 4.2). Remove the disc from its sleeve and insert it into your computer's drive. What happens next depends on what sort of operating system your computer runs.

FIGURE 4.2   The first step to robotics adventure is to install the software!

Installing on a PC

The installation program should begin automatically. If it doesn't run, go to My Computer. Right-click the drive holding the Mindstorms disc and select Autorun from the menu.

> **NOTE**
>
> **Permission Required**
>
> Depending on what version of Windows you are using and the permissions granted to you by the system administrator, Windows might ask you for permission to install the software. If you aren't the administrator, you need to have that person enter his or her password for the installation to continue. If you are the administrator, Windows might still ask if you want to allow the program to make changes. Click Yes and the installation continues.

Installing on a Mac

If you're on a Mac, you may have to follow one semi-tricky set of instructions to get your software to work.

1. Determine the version of your computer's operating system by clicking the Apple menu at the upper left of the screen and choosing About My Mac. A window appears that identifies the version of the OS you are running.

   - If you are running Version 10.5 or lower, you are able to install simply by inserting the Mindstorms disc and launching the install app.

   - If you are running Version 10.6 or higher, you have to follow a slightly different set of directions to install the Mindstorms software.

2. Press Command+Shift+N to create a new folder on your desktop; then drag all the files from the Mindstorms disc into this folder. The files total about 374MB.

3. Locate the file called MindstormsUnivRet.pkg in the Parts directory. Hold down the Control button while clicking on the file (or right-click if you have a two-button mouse); then click Show Package Contents.

4. Open the Contents folder and then the Resources subfolder. Delete the file labeled preflight and then close the package.

5. Click Install.

> **TIP**
>
> **Didn't Delete Preflight File First?**
>
> If you accidentally installed NXT-G before deleting the preflight file, just delete the LEGO Mindstorms NXT folder from your Applications folder and reinstall using the preceding steps.

NXT-G 101

Enough with the preliminaries; let's get started. Are you ready to program? It might seem kind of intimidating at first, but we'll get through it! Let's jump in.

The Programming Block

The central tool of the NXT-G environment is the *block* (see Figure 4.3). A block is a series of programming commands simplified down to one visual cue. It can be tweaked through a series of check boxes to customize its functionality, and the icon changes to show how it has been customized. We get into customization more deeply later.

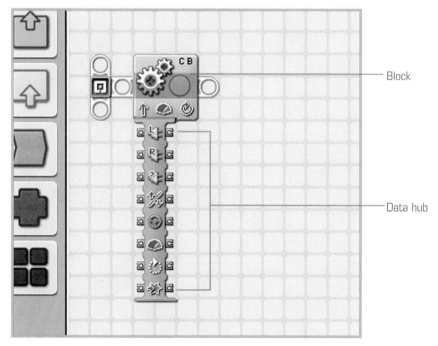

FIGURE 4.3   Blocks embody a string of programming commands.

Blocks have a *data hub* that pops down from the bottom when you click on the tab (see Figure 4.3). The data hub has a number of plugs that you can use to send and receive data and other information from other blocks.

You can even create your own blocks! Say you customize the heck out of a block and want to keep it for later use. You can totally do that.

Figure 4.4 shows the Mindstorms work area, and Table 4.1 explains each component in more detail.

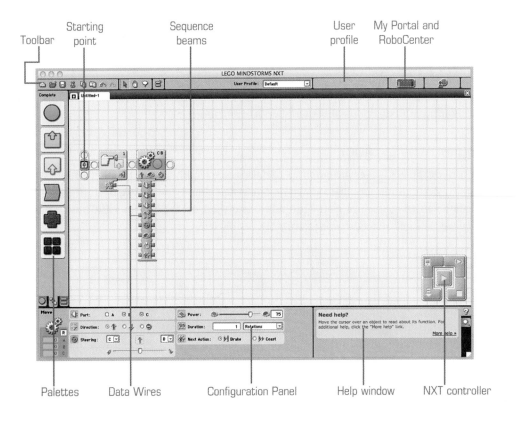

FIGURE 4.4 The Mindstorms work area. This is where the magic happens!

TABLE 4.1 The Mindstorms Work Area

Item	What It Does
1. Starting point	Every Mindstorms program begins at this point. The point is marked with a Mindstorms logo.
2. Palettes	Blocks are organized into palettes—menus that organize the various blocks. There are three: the Common palette features the seven most-commonly used blocks; the Complete palette has everything; and the Custom palette just shows custom blocks created by you or another fan.
3. Sequence beams	These simulated LEGO beams trace the route the program follows as it runs its course. You can split a beam into multiple forks to have more than one operation running at the same time.
4. Data Wires	Wires carry data, text, and commands from one block to another. We don't get into wires much in this chapter, but Chapter 7, "Know Your Sensors," covers them extensively.
5. Help window	This convenient window shows tips about particular elements of NXT-G, which pop up when you hover the mouse pointer over an item.
6. Configuration Panel	Click on a block and see its default setting in this window; then modify it as needed by clicking on the appropriate check boxes.
7. NXT controller	This menu governs the interaction between the computer and a connected NXT brick.
8. Toolbar	The toolbar contains the usual menu options such as opening and saving files, as well as a number of unique tools including utilities for editing images and sounds.
9. User profile	If multiple people use your computer, you can create individual profiles and keep your customized blocks and palettes separate.
10. My Portal and RoboCenter	RoboCenter contains the LEGO models that come with the set. My Portal is a link to LEGO's online Mindstorms resources, including new models, tips, and sample programs.

Commonplace Blocks

Although there are numerous blocks, the following sections discuss a few you're likely to encounter in a typical program.

Color Sensor Block

Mindstorms creations use the color sensor to detect the color of a scanned item and report the result to the NXT brick. This block (see Figure 4.5) controls it! This block is essential for any task that involves the sorting of bricks by color, or for following a colored line.

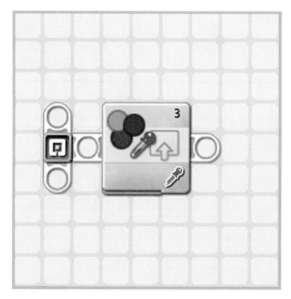

FIGURE 4.5 The Color Sensor block helps your robot to detect the colors around it.

Loop Block

The Loop block causes a sequence of blocks (placed inside it) to repeat until a certain event occurs or the program aborts (see Figure 4.6). This is a great way to keep the size of a program small, because the same blocks are accessed repeatedly.

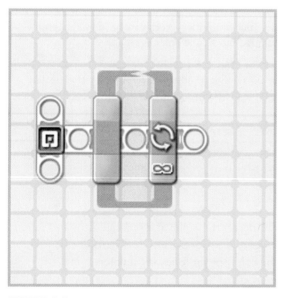

FIGURE 4.6 The loop is one of the classic programming tricks. Just loop it!

Move Block

The default Move block (see Figure 4.7) controls one or more motors. It is used as the basis for many robots because it simplifies certain maneuvers like reverses and turns. You can use the Configuration Panel to modify the block to fit the unique needs of your project.

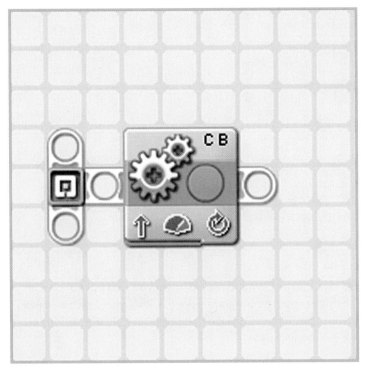

FIGURE 4.7 You use the Move block to control one or more motors.

NXT Buttons Block

The orange and gray buttons on the NXT brick are controlled by the NXT Buttons block, allowing you to interact with your robot's NXT-G program directly from the brick (see Figure 4.8). This is kind of cool because it basically adds "free" buttons to any robot! Of course, you'll need to position the NXT brick in a place where it's easy to access the buttons.

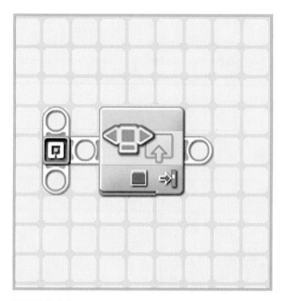

FIGURE 4.8 The NXT brick has buttons built into it. Use them!

Switch Block

What happens when you press that button? What happens when you don't press it? The Switch block (see Figure 4.9) helps control the program's flow as it interacts with a sensor. For example, you could make the robot move in reverse if the ultrasonic sensor detects an obstruction, or go forward if it the coast is clear.

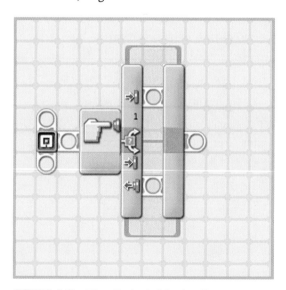

FIGURE 4.9 The Switch block allows you to perform two separate events depending on the input of a sensor.

Ultrasonic Sensor Block

The Ultrasonic Sensor block controls the various functions and options of Mindstorms' ultrasonic sensor, including judging distance and detecting obstacles (see Figure 4.10). It works kind of like a bat's sonar by sending out pulses of inaudible sound and listening for reverberations.

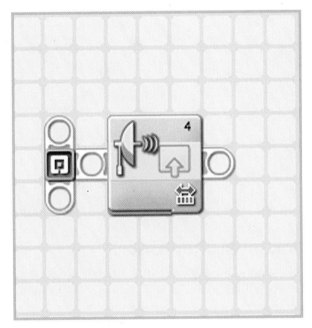

FIGURE 4.10   This block helps your robot see with the ultrasonic sensor.

Wait Block

The Wait block pauses a program for a predetermined length of time or until triggered by another block or a sensor's input (see Figure 4.11). This is great for projects that rely on a lot of sensor input to manage the robot's actions.

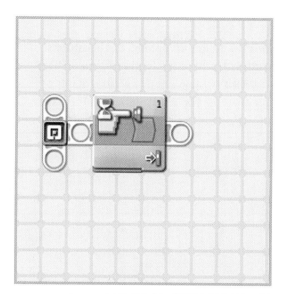

FIGURE 4.11   Wait for it...wait for it!

Programming the Backscratcher Bot

The Backscratcher Bot is fairly simple. No sensors and only one motor make for an elegantly simple program. All we're going to do is program the motor to begin turning with the start of the program and to turn until the button on the NXT brick is pressed.

Create the Program

Let's begin by launching NXT-G (see Figure 4.12). After it is loaded, follow these steps to program the Bot.

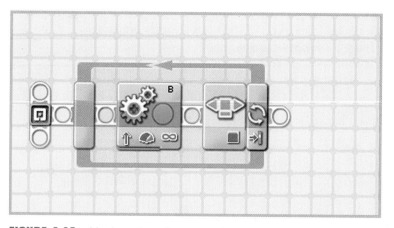

FIGURE 4.12   You've already created the Backscratcher Bot; now let's program it!

1 Drag a Loop block onto the work area.

2 Drag a Move block inside the loop; the Loop block grows as you drop more blocks inside it. Configure it with the correct motor port and change Duration to Unlimited.

3 Click on the right side of the loop to configure it. Change Control to Sensor and then select NXT Buttons from the drop-down menu of sensors. The remaining default settings should be fine.

When completed, the program should look like the one in Figure 4.12.

Connect to the NXT Brick

After your program is complete, it's time to download the program to the NXT brick. But first, you'll have to connect the two! This is how you do it:

1 Plug the NXT into your computer via a USB cable. One is included in the set, or you can use any typical "A-B" cable.

2 Turn on the NXT brick. If you forget, the computer isn't able to see the NXT!

3 Click on the icon on the NXT Controller that looks like a NXT brick.

4 Click Scan in the NXT menu (see Figure 4.13). It takes a few moments for the software to see the brick.

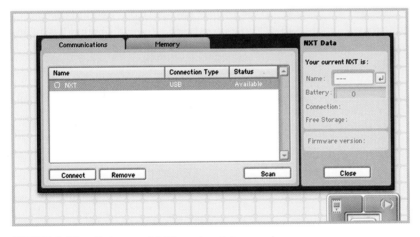

FIGURE 4.13 Select a NXT and get to work.

5 When your NXT appears as one of the visible bricks, select it and click Connect. You're in business!

Download the Program

From the NXT Controller, click the button that looks like an arrow in parentheses. This is Download and Run. Clicking it transfers the program file to the NXT brick and launches it automatically.

TIP

If You Can't Get Your Program to Work

Be sure to read the help files. You can access them through the Help window at the lower-right corner of the screen. Also, don't be hesitant to search the Web for Mindstorms blogs and forums. You'd be surprised how many people have already encountered the same problem!

The Next Chapter

Now that you've learned a little bit about how to program in NXT-G, let's move on to the next model! In Chapter 5, "Project: Clothesline Cruiser," you build and program a robot that can roll along a rope, using its sensors to tell when it has reached the end of the line. And if you're hungering for even more NXT-G know-how, be sure to read Chapter 8, which offers advanced programming tips.

Project: Clothesline Cruiser

Now that we've delved into Mindstorms and its programming environment, let's expand our skills by building and programming an even more complicated robot! The Clothesline Cruiser is a robot that rolls along a clothesline, transporting action figures from one side of the room to the other (see Figure 5.1). One new twist we're exploring with this robot is its interaction with sensors. It has two:

- An ultrasonic sensor that sees with sound, detecting objects in its vicinity
- A touch sensor that sends a signal to the NXT brick when its button is pressed

FIGURE 5.1 The Clothesline Cruiser is a robot that travels along a rope!

Parts You Need

The Cruiser is a bit more complicated than the Backscratcher Bot we built in Chapter 2, "Unboxing the LEGO Mindstorms NXT Set." Gather together the parts shown in Figure 5.2 and follow the steps to build your robot. When you're done, you'll program it. Let's get started!

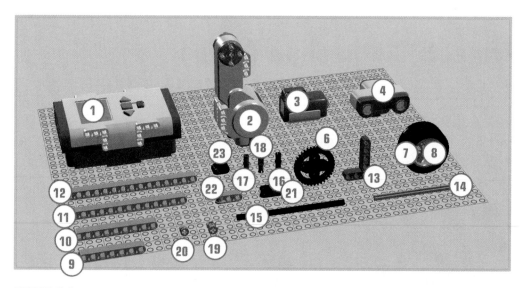

FIGURE 5.2   These are the parts you need to build the Clothesline Cruiser.

1. 1 NXT brick

2. 1 interactive servo motor

3. 1 touch sensor

4. 1 ultrasonic sensor

5. 3 wires (not shown in Figure 5.2)

6. 2 36-tooth gears

7. 4 rims

8. 4 tires

9. 8 7M beams

10. 1 9M beam

11. 6 13M beams

12. 2 15M beams

13. 1 angle beam 3x5

14. 1 9M cross axle

15. 1 12M cross axle

16. 33 connector pegs (black)

17. 15 3M connector pegs (blue)

18. 9 cross connector pegs (blue)

19. 2 bushings

20. 3 half bushings

21. 2 double cross blocks (black)

22. 2 cross blocks (gray)

23. 3 2M beams with cross hole

Step-by-Step Instructions

Let's begin by assembling the Cruiser model.

Follow these steps to build the robot:

STEP 1 Add connector pegs to the NXT brick, eight on each side.

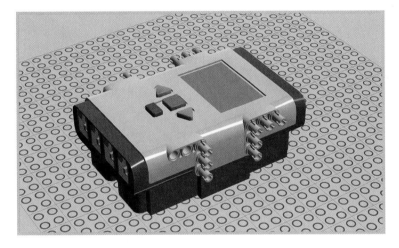

STEP 2   Add five 13M beams as you see here. Note that the NXT is flipped as compared with how it appears in step 1.

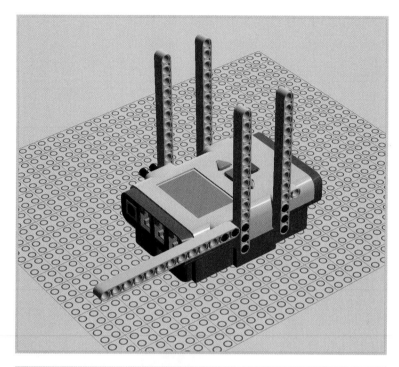

STEP 3   Insert connector pegs on the vertical struts.

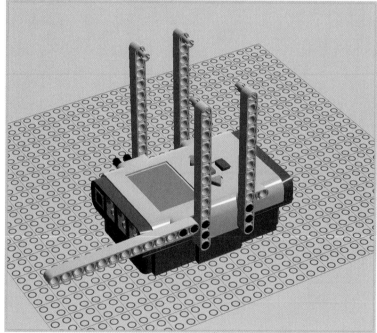

STEP 4 Add two 15M beams to the struts. These beams will support the wheels that enable robot to move along the clothesline.

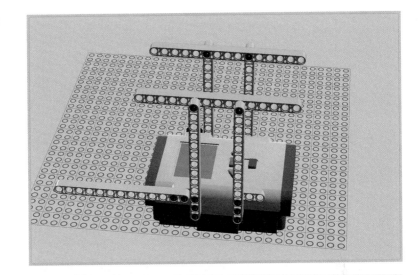

STEP 5 Begin the first wheel assembly by adding a 10M cross axle to a 36-tooth gear. Don't worry too much about getting the axle's placement right; you can adjust it later. The gear doesn't actually work as a gear; it just adds friction to help pull it along the clothesline.

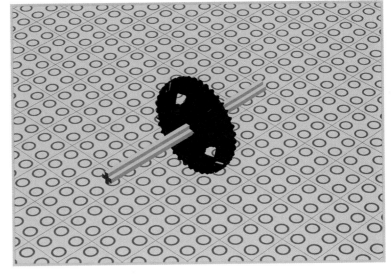

STEP 6 Add rims to the cross axle.

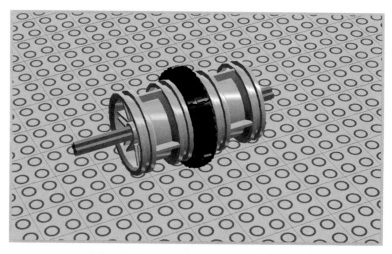

STEP 7 Now add the tires to the rims. They don't really work as tires; they're added to keep the 36-tooth gear positioned on the rope.

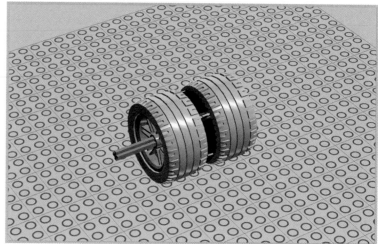

STEP 8 Secure one end with a bushing.

STEP 9 Use a half bushing on the other side. The reason the wheel assembly is positioned off-center is so it balances correctly.

STEP 10 Add the wheel assembly to the main construct.

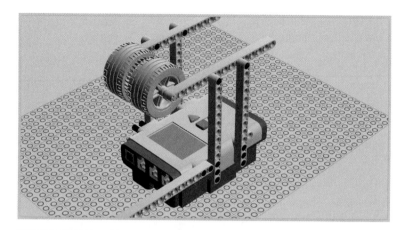

STEP 11 Add a 12M cross axle to another 36-tooth gear. This is the second wheel assembly.

STEP 12   Add rims.

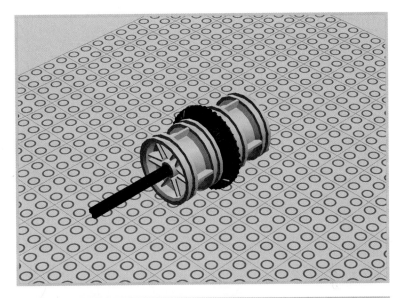

STEP 13   Put the tires on the rims.

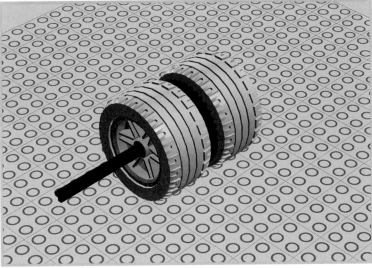

STEP 14 Secure the longer end with a bushing.

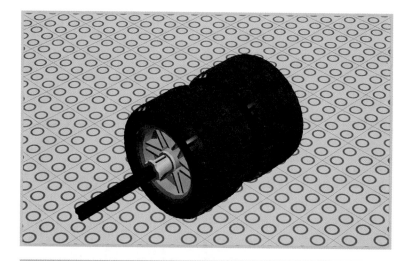

STEP 15 Throw a half bushing on the shorter end.

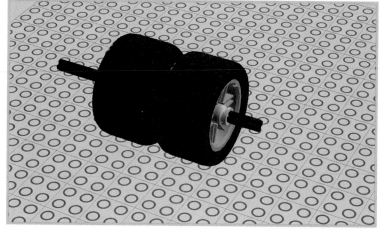

STEP 16 Add the wheel assembly to the robot, keeping the longer length of cross axle to the right as you see here.

STEP 17 Add a half bushing to the short end of the cross axle.

STEP 18   Add a connector peg and a cross connector to a motor.

STEP 19   Add two 2M beams. One goes on the cross connector and is flipped so the peg goes through the cross hole.

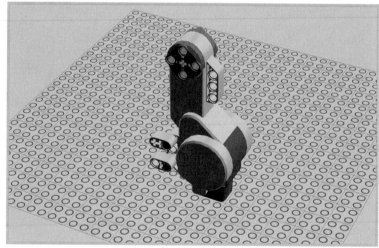

STEP 20 Add a connector peg and a cross connector to the beams.

STEP 21 Add an angle beam to the pegs.

STEP 22   Add connector pegs to the NXT brick. You connect the motor to these pegs.

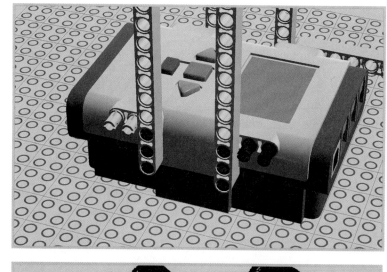

STEP 23   Add the motor assembly to the pegs you just placed, making sure to connect the motor's hub to the cross axle.

STEP 24   Add four connector pegs to the beam jutting off the front of the robot.

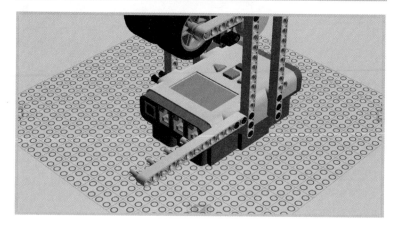

STEP 25 Connect a 7M beam to the pegs. This is the first of several such beams, which comprise the Cruiser's platform.

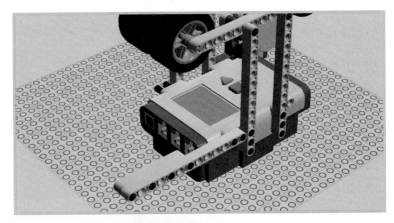

STEP 26 Add three 3M pegs.

STEP 27 Place another 7M beam on the 3M pegs.

STEP 28 And another 7M beam. The platform is starting to take shape!

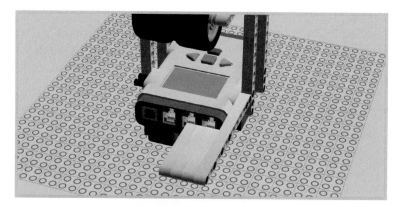

STEP 29 Add two 3M pegs, inserting the long end as you see here.

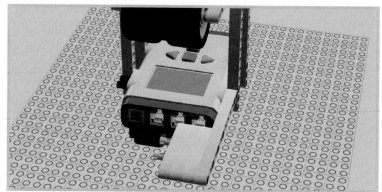

STEP 30 Add another 7M beam. The platform is halfway done.

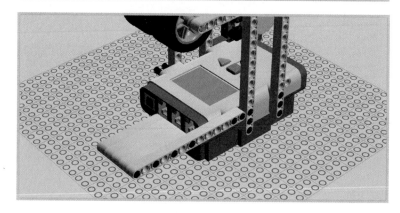

STEP 31 Place a cross connector and three 3M pegs.

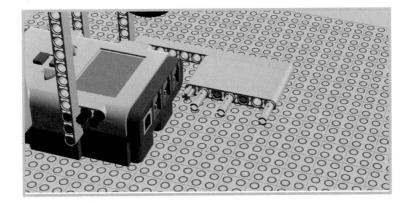

STEP 32 Grab a 9M beam and drop in 4 cross connectors and 3 connector pegs. This is the beginning of the assembly that will support the ultrasonic sensor.

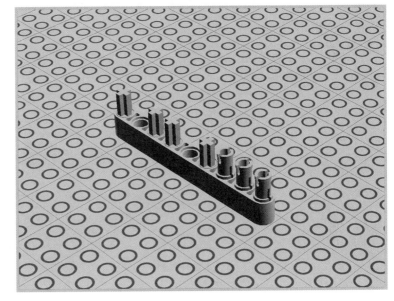

STEP 33 Add two double cross blocks. You'll use the free holes on these elements to connect the ultrasonic sensor to the platform.

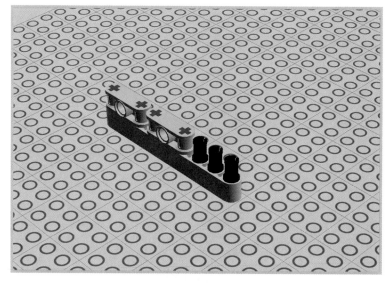

STEP 34 Place the ultrasonic sensor.

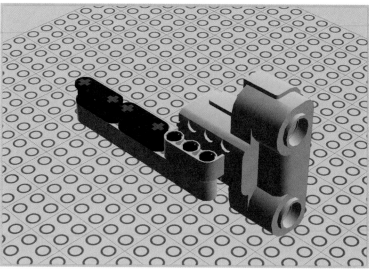

STEP 35 Add the sensor assembly you just created to the robot as you see here.

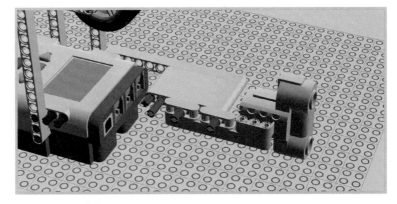

STEP 36 Add a 2M beam.

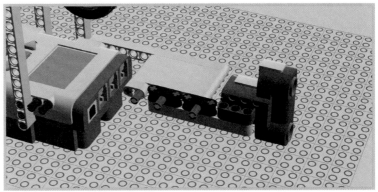

STEP 37 Thought we were done with the 7M beams? Think again! We begin the second half of the platform. Insert the short ends of two 3M connectors. These will support the second half of the platform.

STEP 38 Insert the short ends of two 3M connectors. These will support the second half of the platform.

STEP 39 Add another 7M.

STEP 40 Add a 3M peg, making sure to leave the long end exposed.

STEP 41 Another 7M!

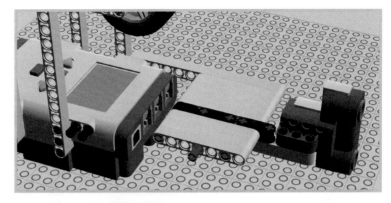

STEP 42 Place two 3M pegs.

STEP 43 Add one last 7M beam! The platform is nearly complete; it just needs some reinforcement.

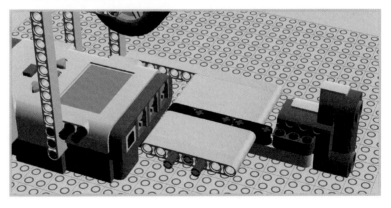

STEP 44 Place two of the regular black connector pegs.

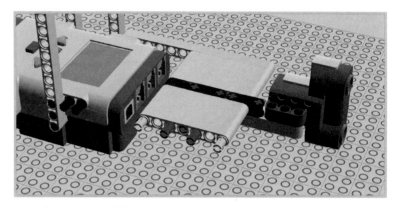

STEP 45 Secure the platform with a 13M beam. We're done with the platform!

STEP 46 Flip over the robot and place two cross connectors on the bottom.

STEP 47 Add 3M pegs to a cross block.

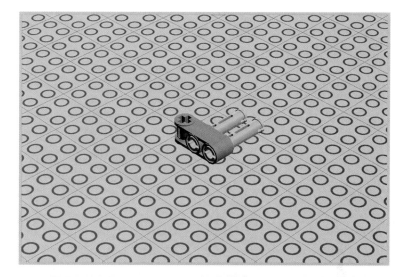

STEP 48 Add the touch sensor.

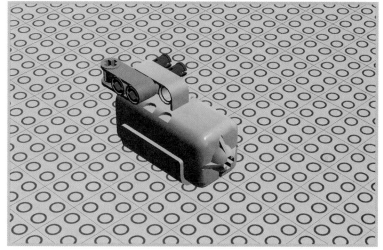

STEP 49 Complete the assembly with a second cross block.

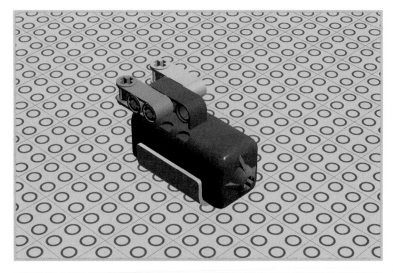

STEP 50 Connect the assembly to the cross connectors on the bottom of the robot.

STEP 51 You're done!

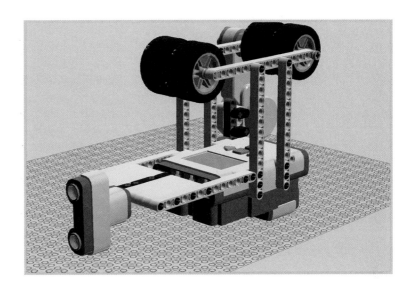

Programming the Clothesline Cruiser

In Chapter 4, "Introduction to Programming," you learned a bit about programming. Now it's time to put that knowledge to the test! Follow these steps to create your program:

1 Launch the Mindstorms software and click on the button next to "Create a new program."

2 Select the Common palette by clicking on the tab in the lower left-hand corner of your window marked with a green circle.

3 Put your mouse pointer on the Loop block in the Common palette. Hold down the mouse button and drag the loop from the palette to the pretend Mindstorms beam (called the Sequence Beam) in the work area.

4 Make sure the loop is selected, then look at the loop's parameters in the bottom of the window. Change Control from Forever to Sensor. A bunch of sensor-specific options should appear.

5 Change "sensor type" to Ultrasonic, on Port 2. Set the distance to 5.

6 Drag a Move block out of the Common palette and drop it right on top of the loop. The Loop block should expand to make room for the Move block.

7 Select the Move block and look at the parameters at the bottom of the window. Configure the Move block to control ports A and B with a power of 100 and Unlimited duration. Direction defaults to Forward, and that's what you want.

8 Drag another Loop block out of the palette and drop it right next to the other one. Set the Control to Sensor—Touch sensor (these are the defaults, and that's what you want). Change Action to Bumped.

9 Drop another Move block into the second loop and configure it the same as the first Move block, except that Direction should be Reverse.

10 Pull a Stop block out of the Flow palette and drop it after the second loop. Your final program should look like the one in Figure 5.3.

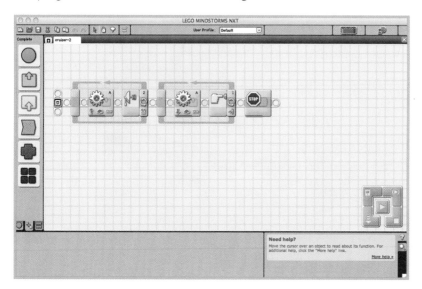

FIGURE 5.3 The NXT-G program showing the operation of the Cruiser.

11 You're done! Now you need to download the program to your NXT brick by connecting the two with a USB cable, selecting the brick from the NXT menu, then clicking on "Download and Run." See "Download the Program" in Chapter 4 if you need a refresher!

Setting Up the Clothesline

The clothesline should be fairly easy to set up. You can use any kind of clothesline or a rope of a similar diameter. It can be stretched between any two reasonably solid objects—even chairs! Try to keep the line as taut and level as you can, although the Cruiser is able to navigate a slightly tilted or bowing clothesline. The length of the rope plays into how much it bows. A 30-foot line will sag in the middle no matter how hard you try to keep it straight. It's just physics; there's nothing you can do about it.

Perhaps the most critical factor is to ensure that the ultrasonic sensor and touch sensor both are triggered correctly. What this means is that each end of the rope needs to have something that a sensor can detect so that it knows to stop the motor at the right position.

For the Ultrasonic, the clothesline's end should have a relatively large and solid surface that the sensor can see. For instance, a chain fence probably wouldn't trigger it. The touch sensor should have an object near the rope that the sensor's button can physically impact. If you're still puzzled, see Chapter 7, "Know Your Sensors" for more tips on sensor use.

What to Do With Your Cruiser?

My original idea for the Cruiser was that it could be used to transport action figures (see Figure 5.4) from one side of the room to another. However, the fact that this is LEGO presents a more intriguing possibility. What if you used the Cruiser as a vehicle for LEGO minifigures?

FIGURE 5.4 A simple elastic band allows you to transport action figures on the Cruiser's platform.

If you want to use your robot to transport LEGO minifigs, swap out one or more of the 7M beams on the platform and replace them with 8-stud Technic bricks. These have seven peg holes, so they can fit in just as easily as those pesky 7M beams yet have studs on top that hold minifigs (see Figure 5.5)!

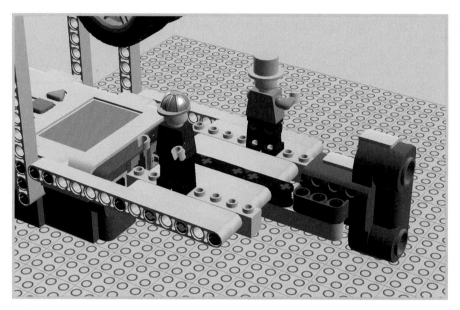

FIGURE 5.5   Action figures are cool, but minifigs are cooler!

The Next Chapter

You've built two models now. Let's ramp up the skill level with some advanced building techniques that you can use in your own projects. Chapter 6, "Building Stronger Models," shows you how to reinforce wobbly robots with angle beams, cross axles, and other techniques.

Building Stronger Models

In this chapter we discuss a number of techniques that you can employ to make your Mindstorms creations stronger and less likely to break apart. Let's face it, LEGO creations are mainly held together by friction, so anything you can do to strengthen your models will help out in the long run. This is especially true in the case of robotics, where the model can be expected to move a great deal more than a static model would. This chapter provides tips on making stronger Mindstorms models.

Use Multiple Pegs

The point of using multiple pegs is obvious: the more pins that connect two beams, the more likely those beams will stick together. Make sure you use the correct pegs, too. Avoid the ones without the friction tabs because they're made to help things move, rather than stay put. You can spot them by the different color. The 2M pegs without friction tabs are usually gray, whereas the 3M pegs without friction tabs are beige. The ones you want are the standard black 2M connector pegs and the blue 3M pegs. They sport friction tabs that help keep the connection from moving around, making your model sturdier.

Connect Each Part to as Many Others as Possible

Similarly, the more parts each element connects to, the stronger the overall model will be. Think of the classic masonry brick pattern (see Figure 6.1) where each brick rests on two others. It's the same principle.

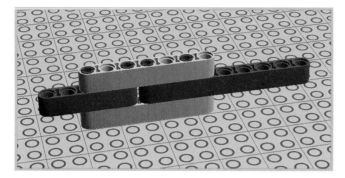

FIGURE 6.1   Securing beams with more beams!

Reinforce Corners with Angle Beams

Angle beams, especially ones with a 90-degree bend, are great for reinforcing the corners of squares (see Figure 6.2). Right off the bat, you're securing it from two different angles, which adds lots of tensile strength. If you flip the L sideways so that one end is sticking up or down, you can secure with vertical beams for added stability.

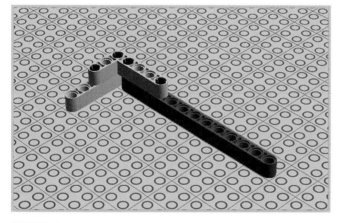

FIGURE 6.2   Angle beams make for strong corners!

Use Combination Parts and Cross Blocks

Combination parts resemble multiple parts that have been merged into one. Take the 3M beam with pegs, for example (see Figure 6.3). At first glance it's the same as a 3M beam with two 3M pegs sticking out of it, but actually it's a lot cooler than that. This beam has holes running perpendicular to the pegs, allowing you to connect to it from two different angles. Even better, it's all fused into a single part, so untoward movement is minimized.

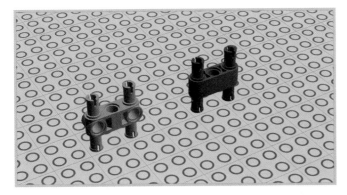

FIGURE 6.3 The 3M beam with pegs is a much more solid connector than the 3M beam with pegs added to it.

Cross blocks work much the same way. For example, the double cross block (see Figure 6.4) has holes pierced in two axes so it can connect to beams to the side as well as above and below.

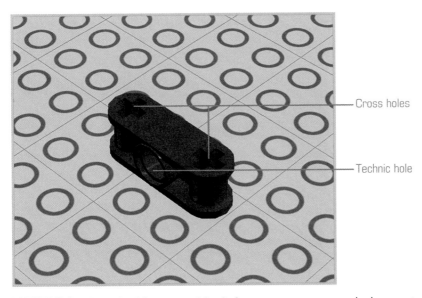

Cross holes

Technic hole

FIGURE 6.4 The double cross block features two cross holes on top, with a technic hole perpendicular.

Attach Cross Axles

Cross axles are your friends. They move your robot when you want them to but keep it still when you need them to. The advantage of an axle is that it can slide through the holes in multiple parts, tying them all together. You can see an example of this in Figure 6.5, where beams are connected with cross axles and bushing separators. If the axles connect to cross holes, rotational movement is restricted because the whole assembly would have to turn.

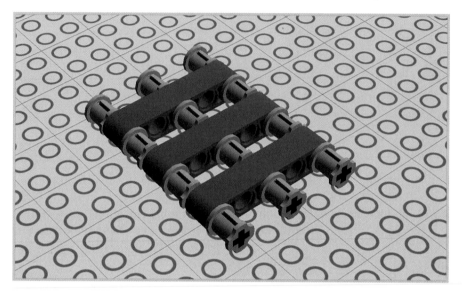

FIGURE 6.5   Cross axles, seen here with bushing separators, secure three beams.

Combine Technic and System Bricks

Wait, what? It's true! You can combine Technic bricks—which are like regular old LEGO bricks pierced with Technic holes—with System bricks. Technic bricks and beams were invented because the studs and tubes connection method for system bricks wasn't strong enough for robots. Technic's pegs and holes method, by contrast, offers a much stronger bond, enabling a robot to move without tearing itself apart. Even better, combining the Technic elements with System bricks creates a much stronger bond. In fact, LEGO's larger model kits often use both types for just that reason.

For example, take the construct shown in Figure 6.6. It consists of a pair of Technic bricks sandwiching two LEGO plates (the thinnest sort of brick, equal to a third the thickness of a regular brick) and secured with a pair of beams. This construct uses both pegs and studs to stay together! This is a super-strong combination that will help prevent your robots from falling apart.

You won't find Technic bricks in your Mindstorms set, so you'll have to buy them online or harvest them from other LEGO sets.

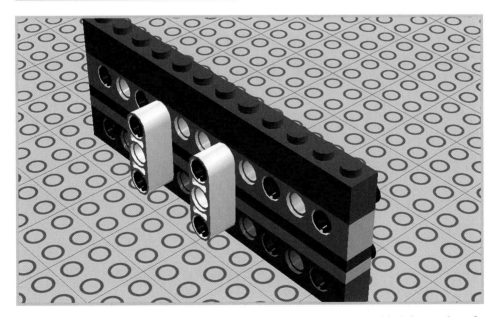

FIGURE 6.6 Stacking Technic bricks with System plates and bricks makes for a strong model.

Use Chassis Bricks

LEGO has a passel of larger bricks (see Figure 6.7) used as chassis structures for the inside of robots. They were designed to make robots stronger, so you might as well make use of them to do just that!

Intriguingly, some of LEGO's less obvious product lines have something to offer. The Bionicle robots often sport a chassis of their own (see Figure 6.8), and most of the parts are Technic-compatible (meaning they will work with your Mindstorms set). Here are links to a small selection of these parts:

- http://peeron.com/inv/parts/64178
- http://peeron.com/inv/parts/32308
- http://peeron.com/inv/parts/64179
- http://peeron.com/inv/parts/53545

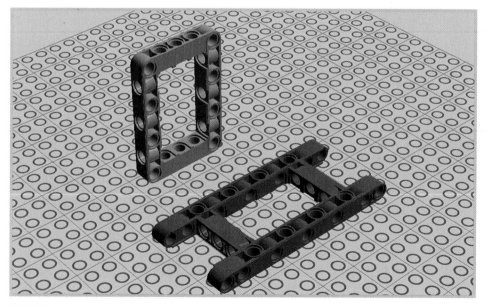

FIGURE 6.7 These chassis elements are just the ticket for reinforcing your model.

FIGURE 6.8 This Bionicle torso is peppered with Technic and cross holes.

The Next Chapter

Next, we tackle those most intriguing and complicated parts in the Mindstorms box, second only to the NXT brick itself. I'm talking about the various sensors that come with the set, as well as those that you can purchase from other companies.

Know Your Sensors

In this chapter, we explore the sensors that come with your Mindstorms set (see Figure 7.1), as well as LEGO-compatible sensors other companies manufacture. Sensors are super important! They're what inform a robot about its surroundings. Being able to detect colors, objects, and movement allows the robot to interact with the environment around it. Even better, those sensors let the robot adjust its program depending on what the robot encounters. How cool is that?

Sound sensor

Touch sensor

Color sensor

Ultrasonic sensor

FIGURE 7.1 How does your robot interact with its surroundings? With sensors.

Mindstorms Sensors

First, let's look at the sensors you get in the Mindstorms set. LEGO chose them because they're the easiest to use and most practical in a basic robot. You'll totally use them throughout your robotics experiments, no matter how far you advance. They're classics!

Touch Sensors

LEGO's Touch sensor is found in many Mindstorms robots. It's basically a button, and pressing it sends a signal to the NXT (see Figure 7.2). That description just scratches the surface, however. First of all, there are three actions you can undertake with just that one button. For instance, the NXT recognizes the difference between a "bump" and a "press."

- A *bump* is a quick press and release of the button (with the pressing lasting less than one second).
- A *press* is when the button is held down for one second or longer.
- The third action, *release*, triggers when the button is pressed (again, longer than one second) and then is allowed to pop out again.

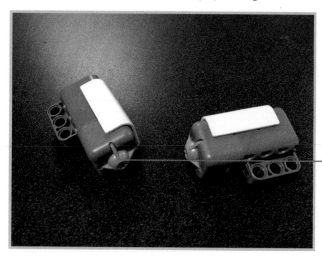

The business end of the Touch sensor—the button

FIGURE 7.2   Touch sensors are an easy addition to any robot.

Touch sensors are one of the most popular sensors that ship with Mindstorms. They can be used as control buttons to allow the user to manually trigger actions in the program. They can also detect obstacles—just add a bumper in front, equipped with Touch sensors! Finally, you can also use Touch sensors to trigger Loop and Switch blocks, making them invaluable for controlling your program's flow.

TIP

Great for Legged Robots!

Touch sensors are also great for legged robots. You can put them on a robot's legs so that it knows when a limb is supporting the robot's weight.

Ultrasonic Sensors

The Ultrasonic sensor (see Figure 7.3) is a great tool for measuring distances. It works by emitting a beam of inaudible sound and listening as that sound bounces back. The time it takes the sound to return tells the NXT brick the object's distance from the Ultrasonic sensor. An example of this would be to program the robot to turn to the right when it got within a certain distance of a wall. The Ultrasonic sensor can be programmed to look for objects at a specific distance, measured as either 0–100 inches or 0–250 centimeters. Objects out of range are ignored.

FIGURE 7.3   Don't let the "eyes" fool you! The Ultrasonic sensor uses sound to "see."

Color Sensors

Although many people call the Ultrasonic sensor the robot's "eyes"—mainly because the sensor actually *looks* like eyes—in some respects the Color sensor (see Figure 7.4) serves the role of an eye a little better. The Color sensor looks for colors within a specific range or outside that range. In other words, you set the program to look for a specific range of colors, and it either reacts when it senses that range or reacts when it detects anything *but* that range. You can also set the sensor to work as a light sensor from within the Color sensor block that we discussed in Chapter 4, "Introduction to Programming." That means it looks for either any old light or a specific color of light with intensity set using a slider that ranges from 0 to 100. Finally, and even more intriguingly (for me), the Color sensor also can work as a color lamp. It uses the Color Lamp block, found in the Action menu of the Complete palette, and the color it emits can be either red, green, or blue. All in all, a very useful sensor!

FIGURE 7.4   The Color sensor also detects colors and works as a lamp!

Sound Sensor

LEGO Mindstorms NXT 2.0 doesn't actually ship with a sound sensor, although the previous edition did (see Figure 7.5). However, there are tons of these sensors floating around, and they are still supported by LEGO's hardware and software, making them a cinch to add to a robot. When set to listen, this sensor returns a value from 1 to 100 depending on the volume of the sound detected. You can set a threshold in your program that triggers an action when the sound reaches a certain level. For instance, a sound sensor would be useful with a robot that moves only when a loud noise occurs. Want to buy one? They're available on legoeducation.com.

FIGURE 7.5 LEGO's sound sensor gives robots the ability to react to noises. Unfortunately, it's not available in the standard Mindstorms sets.

Motor

Wait, what? How can a motor be a sensor? The answer is that the Mindstorms motors pack encoders—little slotted disks—that keep the NXT informed as to how far the motor's hub has turned (see Figure 7.6). Ordinarily, this information is used to track how far the robot has rolled or turned. For instance, you can set the robot to stop after it travels a specific distance. However, the motor can also be used as an input device. Say you want to steer your robot with a remote control. You can accomplish this task easily with the help of a motor connected to your robot with a long wire. You turn the hub in the direction you want, and the NXT-G program takes the data from the motor and uses it to steer the robot. You can also make a wireless remote control if you have two NXT bricks. Simply connect the two via Bluetooth. The remote control is basically a NXT with a motor attached to it.

FIGURE 7.6   LEGO's motors are more than motors—they're sensors!

Calibrating Sensors

LEGO sensors typically return a reading from 0 to 100, but what exactly is 0 and what is 100? For sensors to perform as you expect, you might need to calibrate them. *Calibration* is the process of setting the parameters of the sensor so it knows the minimum and maximum it will be looking for. For instance, an uncalibrated light sensor might think that a very bright light bulb is only an 80, so it keeps expecting something brighter. Conversely, a robot in a dark room might only return readings of 20 and under, hobbling the robot's performance.

There are two ways to calibrate a sensor. First, the Mindstorms software has a menu option under Tools. To use it, plug in a Light (Color) sensor or Sound sensor to a powered-up NXT brick that is connected to your computer. When you select the Calibrate Sensor option under Tools, you see the available options, which for a default-configured NXT includes Light and Sound sensors (see Figure 7.7). Some sensors don't need to be calibrated, whereas others that do are supported only by third-party tools. For instance, the Compass sensor discussed later on in this chapter must be calibrated to work, and it uses a custom calibration block that can be downloaded from the same site as the product itself.

The other way to calibrate a light or sound sensor is to add a calibration block to a program via the Advanced menu. The advantage to calibrating in-program is that you can adapt to whatever conditions the robot finds itself in. For instance, if it recalibrates every time it runs, the robot would work equally well in a brightly lit room or a dark room.

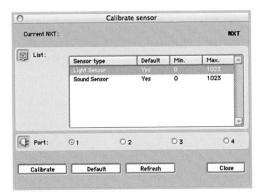

FIGURE 7.7 Want to calibrate your sensors? The Calibrate Sensor menu option is for you.

TIP

Adding a Calibration Block

One way to calibrate a light or sound sensor is to add a calibration block to a program. You'll find it in the the Advanced menu. Having a calibration block in your program allows the robot to recalibrate on the fly, responding to different light or sound conditions. For example, in a quiet house the robot might need a much different set of parameters for detecting sound levels than a noisy backyard.

Third-Party Sensors

Although Mindstorms covers all the obvious bases, there are actually a lot of intriguing alternatives available, created by small businesses such as Dexter Industries (dexterindustries.com), Mindsensors (mindsensors.com), and HiTechnic (hitechnic.com).

NOTE

More Where These Came From

The third-party sensors discussed here just scratch the surface. There are many others, such as GPS receivers, pressure sensors, and thermometers. If it's electronic, it can probably be made into a Mindstorms sensor, and chances are, someone has done it!

Compass Sensor

The compass sensor, offered by Dexter Industries, tells the NXT which direction the sensor is pointing. It homes in on magnetic north, but this makes the sensor subject to errors when close to a magnet or an electromagnetic source such as a motor. You can see the compass sensor in Figure 7.8.

FIGURE 7.8   Dexter Industries' compass sensor tells you what direction is north.

Passive Infrared (PIR) Sensor

The Passive Infrared (PIR) sensor, available from HiTechnic, works much the same way as the ultrasound, except that it uses invisible light called *infrared* to detect motion. It's not so good at detecting inanimate objects, but it excels at sensing the changes of temperature caused by a person walking by.

Wi-Fi Sensor

The Dexter Industries Wi-Fi Sensor shown in Figure 7.9 plugs your robot into a nearby Wi-Fi network, allowing it to interact with the Internet! You could, for example, set the NXT to send out a tweet whenever your robot performs a certain action.

FIGURE 7.9   The Wi-Fi Sensor connects your robot to the Internet. Credit: Dexter Industries

Magnetic Sensor

The HiTechnic Magnetic Sensor detects the presence of a magnetic source nearby, allowing you to trigger an event in your program when a magnet passes nearby. An example of this would be to monitor the movement of a robot around obstacles. By placing magnets at key intervals, you could tell the robot when it's approaching the next obstruction.

Flex Sensor

A Flex sensor detects when a flexible element bends (see Figure 7.10). This capability has intriguing possibilities! For instance, you could sew the Flex sensor into a glove and then build a robot that closes its claw when you make a fist. How cool would that be?

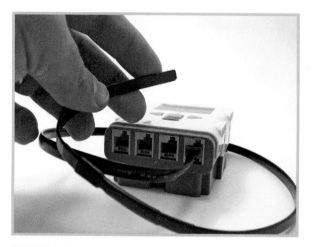

FIGURE 7.10   The Flex sensor can tell when you bend it. Credit: Dexter Industries

Voltage Sensor

Mindsensors' VoltMeter detects the voltage of a circuit connected to the sensor's leads. You could use this capability to make a simple LEGO battery checker, for example, allowing you to check all those old batteries lying around to see whether they're still good.

Barometric Sensor

Another HiTechnic sensor, the Barometric sensor detects both temperature and barometric changes. It can sense air pressure changes of up to 1/1000 of an inch of mercury as well as temperatures to 1/10 of a degree of Celsius, making it ideal for Mindstorms-based weather monitoring.

Inertial Motion Sensor

Dexter Industries' dIMU, an Inertial Motion sensor, measures acceleration, tilt, and rotation on three axes, allowing your robot to know exactly how it is positioned and how fast it's accelerating (see Figure 7.11).

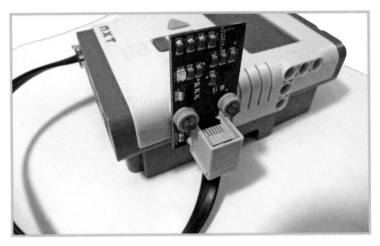

FIGURE 7.11   The dIMU sensor informs the NXT brick about the robot's movements. Credit: Dexter Industries

The Next Chapter

In Chapter 8, "Advanced Programming," we learn advanced programming techniques! We explore all the cool things you can do with data wires within your NXT-G program, and we check out variables, datalogging, and even how to create your own programming block!

Advanced Programming

In this chapter, we learn a bit more about Mindstorms programming and how it works. First and foremost, we explore *data wires*, NXT-G's on-screen representation of the flow of data (see Figure 8.1). When one programming block needs to send information to another block, it uses data wires.

Wires are the key to turning an otherwise blah program into a sophisticated and responsive one that can pull in data from its sensors and use that data to adjust the behaviors of various blocks. Learning about wires is one of the keys to advanced programming!

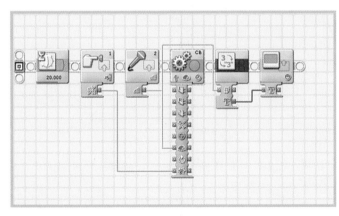

FIGURE 8.1 Look complicated? Let's learn more about programming.

Data Wires

In the Mindstorms programming world, wires regulate all the flow of data throughout the program. What if you want to have the reading from the Light sensor regulate how far your robot travels? Wires send data from one block to the other and make it happen. There are four kinds of data wires, described next.

Green Wires

The green wires carry Logic data, which is Mindstorms parlance for things that are either true or false. Note that some blocks refer to "true" or "false" as "yes" or "no," but for purposes of simplicity we're just going to use "true" or "false." Put simply, there are literally only two possible types of data transmitted by a Logic wire: absolutely true and absolutely false. Here's an example: A button is either pressed or not pressed. If it's pressed, the program returns a true response; if it's not pressed, the result is false. We see this in action in Figure 8.2.

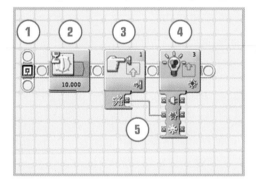

FIGURE 8.2   Green wires carry Logic data, which represent a statement that can be either true/yes or false/no.

1. The program begins with the starting point. Every program begins there and works its way to the right, following each block in sequence.

2. The Wait block pauses the program for 10 seconds. After that time has elapsed, the program continues.

3. The Touch Sensor block is the next in line. If the button is pressed when this block executes, a Logic signal is sent along a green data wire to the Color Lamp block.

4. The Color Lamp block turns the color sensor into a light. The green Logic wire delivers either a true or false signal. In the former case, the light comes on! If the button isn't pressed, a false signal is sent and the light remains dark.

5. This marks the path of the green data wire as it travels from the Logic plug of the Touch Sensor block's data hub to the Color Lamp's hub.

Yellow Wires

Yellow wires carry *Number* data. What the heck is this? Is Number data any old number? Well, sort of. It's actually kind of tricky. Let's break it down:

- An Ultrasonic sensor's reading is a Number because the sensor generates a number based on the robot's distance to an object.
- A Touch sensor's output isn't a Number because it creates only a yes or no, which is called Logic data in the Mindstorms world.
- Number data can be manipulated with math and can change a motor's or sensor's settings. In Figure 8.3, a Rotation Sensor block returns Number data showing the degrees of rotation, and sends that data along a yellow data wire to the Motor block, which alter's the motor's duration (how long it turns) accordingly.

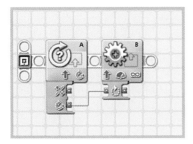

FIGURE 8.3 Yellow wires send Number data, which is just what it sounds like!

Orange Wires

Text data is transmitted through orange wires. Text differs from Number data in that it's an actual string of text, such as a sentence or name. An example of this could be the output of a Text block, which can't output Number or Logic data. You could spell out a sentence with Text data, but you couldn't do math with it or make the motors turn a certain number of rotations.

Text data is unique in that it is the only kind of data that can be displayed on the NXT brick's screen. If you wanted to display a sensor's reading, you'd first have to convert the reading from Number data to Text data using a special block—the Number to Text block. Figure 8.4 shows an example of how to use the Number to Text block.

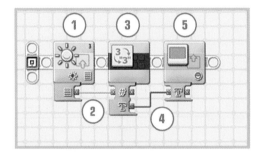

FIGURE 8.4   Text data, carried by orange wires, consists of strings of letters, numbers, and punctuation marks.

1. For example, look at the reading from your Light Sensor. The reading is output through a yellow wire as Number data.

2. The yellow wire begins with the Intensity port of the Light Sensor block's data hub and travels to the number port of the Number to Text block.

3. The Number to Text block converts Number data to Text data, not surprisingly! Data converted this way can then be displayed on the NXT brick's screen.

4. The orange Text wire travels from the Text output plug of the Number to Text block to the input data plug of the Display block.

5. The Display block accepts the Text data and displays it on the screen!

Gray Wires

If you see a gray wire on the NXT screen, that means the wire is busted and won't transmit data. There can be a number of reasons why the wire isn't working. If you try to connect two mismatched plugs—a Number plug and a Text plug, for instance—the wire will turn gray, indicating that the connection isn't going to work. Another example might be a broken string of wires where data isn't supplied to one block, so it can't transmit the data to the next. In Figure 8.5, we see a wire string between two output Logic plugs of two sensors. Because two output plugs can't be connected, a gray wire is the result. Gray wires in your program will result in a failed download, so you must make sure your wires are in order if you want your robot to work.

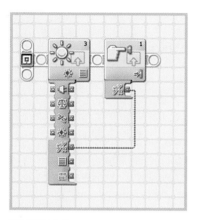

FIGURE 8.5 Uh oh! A gray wire indicates a broken connection.

Connecting Wires

The first step to connecting two blocks with wires is to open their data hubs. Some blocks automatically pop out their hubs when placed on a sequence beam. The tricky thing is that they don't always pop out all the way. If you don't see the plug you want, click on the tab to extend it all the way. Clicking on the tab again after you've added wires will hide all the unused plugs.

Next, click on the output data plug of the type of connection you want to make. Remember, inputs are on the left, and outputs are on the right. Then, while holding down the mouse button, "draw" a line to the input plug of the block you want to connect to (see Figure 8.6). If you did it right, the wire will be colored orange, green, or yellow. If you messed it up, the wire will be gray. If you change your mind and want to delete the wire, click on the input plug the wire leads to.

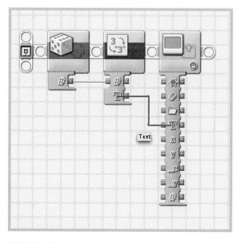

FIGURE 8.6 Want to connect two blocks with a wire? Just click and draw it with your mouse pointer.

Additional Blocks

Although we covered a bunch of blocks in previous chapters, there are a few more yet to discuss. You won't use these blocks in every program, but you'll be glad to have access to them when you finally need them!

Variable Block

In the programming world, a variable is a piece of data that can be modified during the operation of the program (see Figure 8.7). For instance, a Light sensor could tell you whether the room was light or dark, and the result would be saved within the Variable block so that the robot would always know. Variables can be either Logic, Number, or Text, and the block can be set to read or write the variable.

When modifying a variable, you set the block to "write." When you want to access the value stored, you set the block to read and whatever data is stored may be accessed by another block.

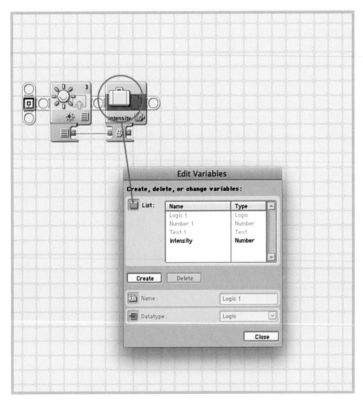

FIGURE 8.7   The suitcase icon represents a variable—in this case, a Number variable.

Constant Block

The Constant block creates a Number, string of Text, or a true/false Logic statement that when placed into a program, never changes. That constant can be accessed through data wires as needed and can be given its own name so you can easily reference it.

You use constants much the same way as you'd use a variable, except that once defined, a constant can't be changed during the course of the program. In the example in Figure 8.8, a Logic constant activates a light.

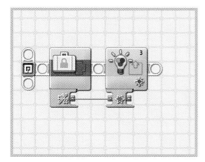

FIGURE 8.8   Unlike a variable, a constant never changes.

CREATING A VARIABLE OR CONSTANT

Variable and Constant blocks are a little unusual in the Mindstorms world because you may adjust their settings through the Edit, Define Variables (or Constants) menu options at the top of your screen, as well as being modifiable at the block level.

As shown in Figure 8.7, you can use the Edit menu to create and name a variable and decide what sort of data it will store. When you configure a Variable block, your new variable (in this case, "intensity") will appear as one of the options for that block. Defining constants works much the same way except that once set, the constant cannot be changed, hence the padlock logo on the icon.

Note that creating a variable or constant through the Edit menu will generate global values that can be used throughout the program. For example, if you set a constant of Light? = True using the Edit menu, then any Constant block will have Light? as one of the choices. Creating or modifying a variable through a block's settings will affect only that one block.

Random Block

A Random block generates a random number between 0 and 100 (the default) or any other range smaller than that, every time you launch the program. For instance, you could make the range 1–20 if you wanted, or 15–75. In the example shown in Figure 8.9, a Random block generates a number, which sets the duration of a Motor block. Changing the range of the random number ensures that the motor doesn't roll too far!

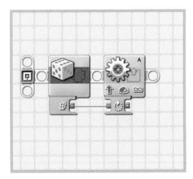

FIGURE 8.9 A Random block creates a random number that can be used to add the unpredictable to your program.

Keep Alive Block

Although normally you can set the amount of time before the NXT brick falls asleep directly from the NXT menus, the Keep Alive block lets you renew that timer. For instance, you could have the brick stay awake during a longer program and then fall asleep when done.

If you select the Keep Alive block, you'll immediately notice that the configuration panel is black; there are no settings to change. The way it works is that whenever the program launches the Keep Alive block, the NXT's built-in sleep timer resets. So, if you set the NXT to fall asleep after five minutes, then every time the Keep Alive block is run, that five-minute timer starts over at the beginning. In other words, if you want to keep your program running indefinitely, include a Keep Alive block in a loop as shown in Figure 8.10.

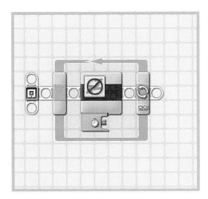

FIGURE 8.10   The Keep Alive block inside a loop keeps your NXT brick awake.

Light Sensor Block

The Light Sensor block uses a Light sensor to detect ambient light levels, sending either a numeric light reading or a true/false via a data wire. As shown in Figure 8.11, the Light sensor returns a "true" value" if reading is a 50 or higher. The Logic wire connecting the Light Sensor block to a Motor block activates the motor's power and it starts moving!

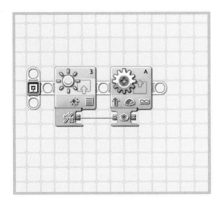

FIGURE 8.11   The Light Sensor block takes information from the sensor and shares it with other program blocks.

Rotation Sensor Block

The Rotation Sensor block takes data from a motor's encoder—the widget inside the motor that tells the NXT how far and how fast the hub turns—and returns either the number of rotations or the number of degrees made by the motor to the program, which compares it against a preset value. The example in Figure 8.12 shows how you could control the turns of a motor's hub with another motor; the block detects the rotation of the first motor and instructs the second motor to mirror it.

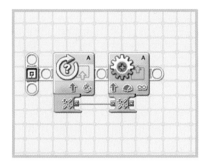

FIGURE 8.12   How far did that motor hub turn? The Rotation Sensor block can tell you.

Display Block

The Display block controls the NXT brick's screen, allowing you to use the display as part of your robot. You can set it to show text, or you can also have it display a drawing like a smily face. Figure 8.13 shows how you could set the Display block to display the reading of an Ultrasonic sensor, using the Number to Text block to convert the Number data the sensor outputs into the Text data the Display block can handle.

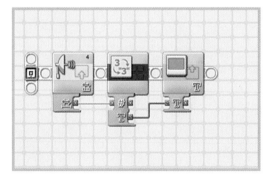

FIGURE 8.13   The Display block controls the NXT brick's LCD screen.

Bluetooth Block

Multiple Bluetooth blocks manage the wireless communications of your robot. The Bluetooth Connection block (the leftmost block in Figure 8.14) activates a wireless connection between two NXT bricks, providing the two have previously connected and appear on each other's contact list. The Send Message (middle block) and Receive Message (rightmost) blocks control the stream of data—Logic, Text, or Number—that can be transmitted between NXT bricks via Bluetooth. Note that that Figure 8.14 doesn't actually show a viable program; I just put all three blocks on a beam so you could see them.

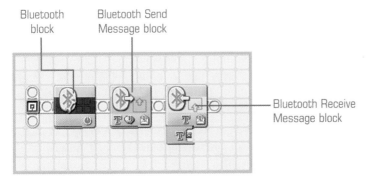

FIGURE 8.14 Mindstorms' Bluetooth capability is controlled by these three blocks.

Logic Block

As we've already discussed, Logic data can be either true or false. The Logic block manages inputs from up to two sources of Logic data and then sends out a Logic signal to the next block. In the case of Figure 8.15, two touch sensors trigger a Color Lamp block; the Logic block looks for *both* buttons to be pressed simultaneously before it triggers the lamp.

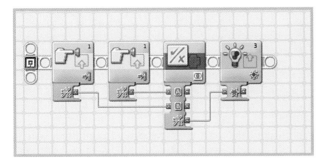

FIGURE 8.15 The Logic block accepts data from two input plugs and outputs a yes/no.

TIP

Logic from Two Places

The Logic block works with the Mindstorms Logic function, which governs conditions that are either true or false. The cool thing is that it can accept Logic data from two data wires, allowing it to reference both when making a decision. For instance, it could look to see if Data A is true while Data B is false, or both are true, or both are false.

Creating Your Own Blocks

One of the most intriguing aspects of NXT-G programming is the possibility of combining one or more blocks into a single customized block. The way this works is that you can grab a series of programming blocks and their data wires and combine their functionality into a single block, called a My Block. If you're planning on using this new block a lot, you can even save it and drop it into all of your programs.

Here's how it works: Say you want to display your light sensor readings on the NXT brick's screen, which would ordinarily take three blocks—a Light Sensor block sending data to a Number to Text block and then to a Display block. The Number to Text block takes the Number data from the sensor and converts it to Text; the Display block then shows the text on its screen. By combining these three blocks into a single My Block that has already been configured the way you like, you can save yourself a lot of time!

Here's how you do it:

1. **Lay Out Your Blocks**—First, arrange the blocks just as you would do in a normal Mindstorms program with all the data wires set up the way you want. Click and drag to select all the blocks you want to turn into a My Block (see Figure 8.16). It doesn't have to be every block in the program!

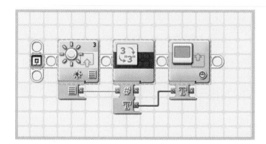

FIGURE 8.16   To create a My Block, first start with the snippet of program you want to turn into a block.

2. **Launch *Create My Block***—Next, select Create My Block from the Edit menu. Your blocks appear in a pop-up window labeled My Block Builder (see Figure 8.17). This view is completely editable, so if you need to make changes, or you accidentally added on or left off a block, you can fix the problem in this view. Now you can edit your My Block any time you want! Note that editing a My Block edits it for all uses of that block, so if you edit it to draw from Port 1 instead of Port 3, forever after that block will pull data from Port 1. If you want to pull data from two sensors, you need two separate versions of the My Block. You'll also need to come up with a name for your block; choose one that describes precisely what the block does.

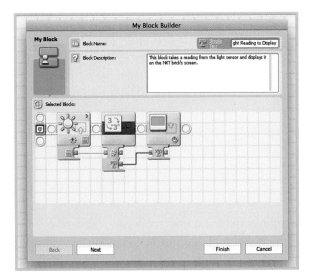

FIGURE 8.17 Launching the Create My Block option brings your blocks into a separate window.

3. **Edit the Icon**—After you set the custom block's program, click on Next and you are able to edit its icon (see Figure 8.18). This simple process involves adding one of a wide assortment of icons to a teal-colored block. When you're done, click Finish and you return to your original program (see Figure 8.19), but with your new block in place of the code you selected. It's slick!

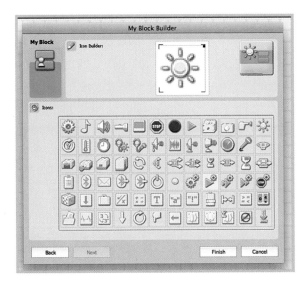

FIGURE 8.18 You can customize your icon to make your block more recognizable.

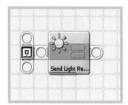

FIGURE 8.19   The end result of the process is a custom block that appears in your program!

The Next Chapter

In Chapter 9, "Project: Rebounder," we build our third and final robot, the Rebounder. This cool tank-treaded roller uses a couple of cool programming tricks that will keep you thinking about NXT-G! Even better, you get to create a robot that interacts with its environment.

Project: Rebounder

Our third and final robot is the Rebounder, a vehicle that rolls around on tank treads and bumps into everything in sight (see Figure 9.1). The good news is that we'll program the robot to react to these obstacles. When it hits something, the Rebounder will choose a random direction and roll off in search of something else to hit. Let's get building!

FIGURE 9.1 This robot exists to bounce off the walls!

Parts You Need

Here are the parts you need to build the Rebounder:

NOTE

Not All Parts Shown

Figure 9.2 shows most of the parts you need for the Rebounder but not all. In addition to the parts shown here, you need four **wires (whatever lengths work for you)** and two tank treads.

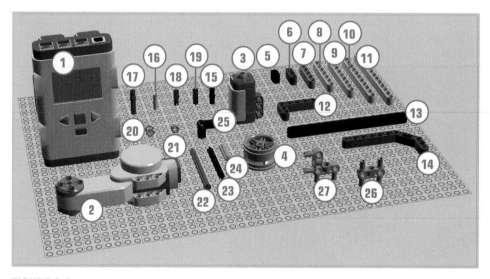

FIGURE 9.2 Grab these parts to make the Rebounder.

1. 1 NXT Brick
2. 2 Interactive Servo Motors
3. 2 Touch Sensors
4. 4 Rims
5. 2 2M Beams with Cross Hole
6. 2 3M Beams
7. 4 5M Beams
8. 1 7M Beam
9. 1 9M Beam
10. 1 11M Beam

11. 3 13M Beams
12. 2 Angle Beams 3×5
13. 2 15M Beams
14. 2 Double Angle Beams 3×7
15. 20 Connector Pegs (black)
16. 2 Connector Pegs (gray with no friction tabs)
17. 8 3M Connector Pegs (blue)
18. 6 Cross Connector Pegs (blue)
19. 2 2M Cross Axles (red)

20. 8 Bushings
21. 4 Half Bushings
22. 2 3M Cross Axles
23. 2 6M Cross Axles
24. 2 8M Cross Axles with End Stop
25. 4 90-Degree Angle Elements
26. 6 3M Beams with Pegs
27. 2 Angle Beams with Pegs

Step-by-Step Instructions

The Rebounder isn't a terribly difficult build, so you'll do just fine. That said, there are a couple of tricky steps in the build instructions, but I guide you through all right!

TIP

Look for the Blue Pieces

Remember, as I add new pieces to a model, I show them in the associated figure in blue.

STEP 1   Let's start off by working on the Rebounder's bumper. Grab a 13M beam and add a peg to each end. Use the gray connector pegs with no friction tabs.

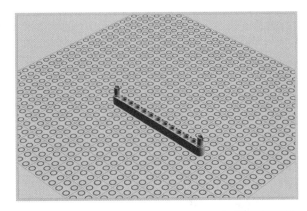

STEP 2   Add two angle beams. They will rotate freely, so be warned!

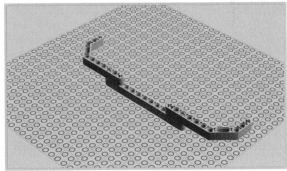

STEP 3   Add two blue cross pegs and two red 2M cross axles.

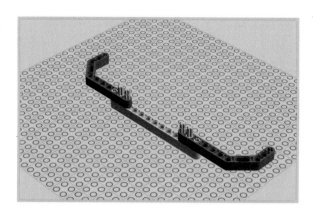

STEP 4   Now let's work on a different part of the bumper. Add a black peg to one of these angled axle connectors.

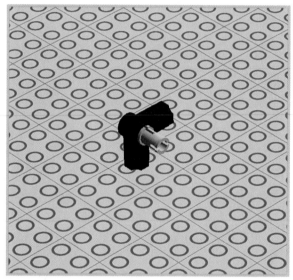

STEP 5   Add another axle connector! Now make another assembly just like you see in the figure.

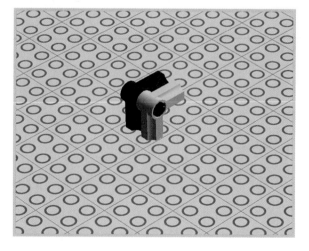

STEP 6   Add the two assemblies to the bumper.

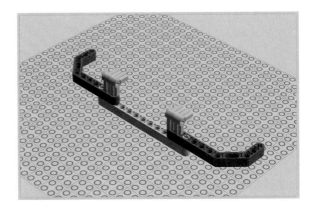

STEP 7   Add two blue cross pegs.

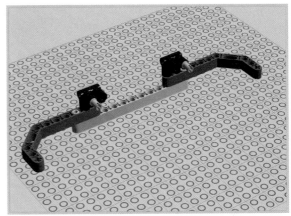

STEP 8   Add two 3M cross axles. The bumper will connect to the robot's body with these axles.

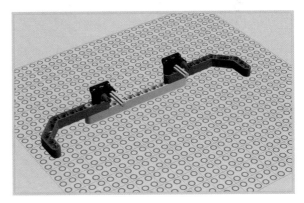

STEP 9   Connect an 11M beam. The bumper is done!

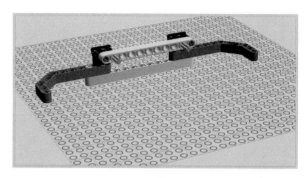

STEP 10   Let's work on the main car, starting with a motor. Add a 3M beam with connectors.

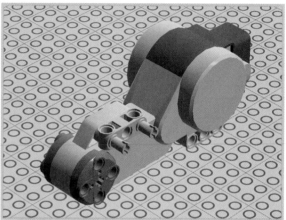

STEP 11   Add a 5M beam.

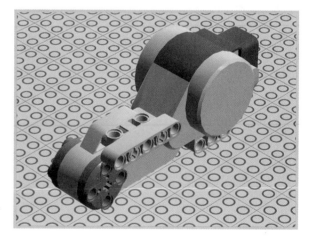

STEP 12 Add three 3M connector pegs.

STEP 13 Connect two more 5M beams.

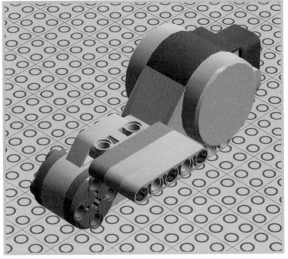

STEP 14   Add another 3M beam with pegs.

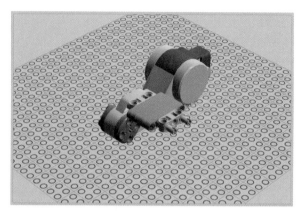

STEP 15   Now let's add another motor! It's kind of starting to look like a car already!

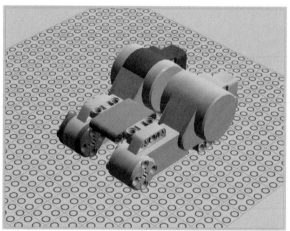

STEP 16   Add four black connector pegs.

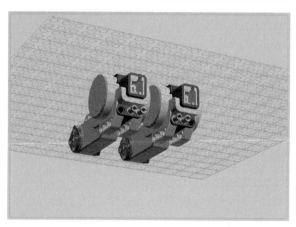

STEP 17   Connect a 13M beam.

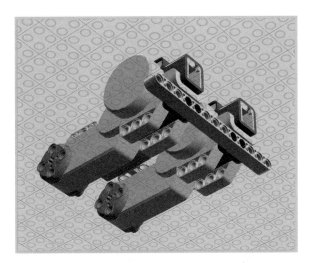

STEP 18   Attach two of these angle beams with pegs.

STEP 19   Add two 15M beams.

STEP 20　Slide through a 6M cross axle and add a bushing on one end and a half bushing on the other. One of the rear wheels will mount on this axle.

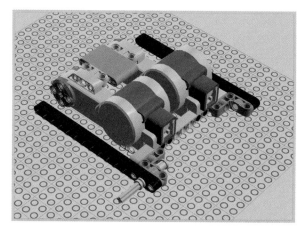

STEP 21　Repeat on the other side! Obviously, this assembly will hold the other rear wheel.

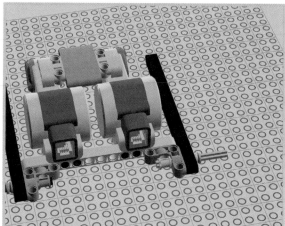

STEP 22　Slide an 8M cross axle with end stop through a rim.

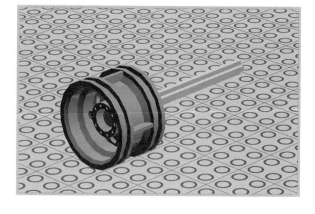

STEP 23   Secure it on the other side with a half bushing. Now make another one just like this!

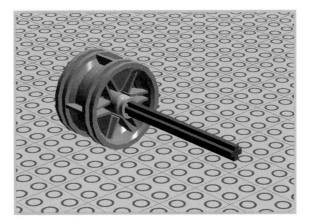

STEP 24   Slide the cross axles partway through the beams. You might think they'll fall out, but don't worry; we secure them in the next step.

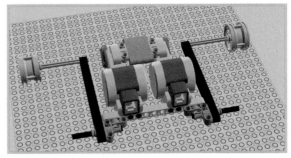

STEP 25   Insert the cross axles through two bushings each and then into the motors' hubs. Notice a little bit of each axle sticks out the opposite side of each motor.

STEP 26 Add rims to the other axles.

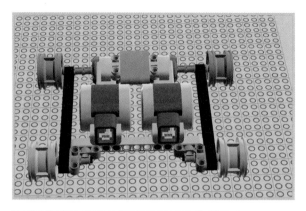

STEP 27 Secure one wheel with a bushing.

STEP 28 Add a bushing to the other rim.

STEP 29 Add two 3M beams with pegs. The touch sensors will mount to these parts.

STEP 30 Attach the Touch sensors! Make certain the Touch sensor buttons face away from the rear of the motors.

STEP 31 Add two connector pegs.

STEP 32 Throw on a 5M beam. The NXT brick will connect to this beam.

STEP 33 Add some more connector pegs, six in all.

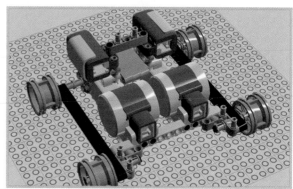

STEP 34 Connect angle beams onto the rear of the robot.

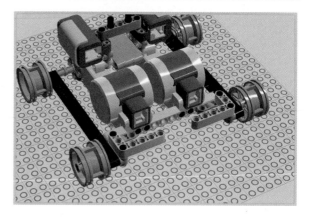

STEP 35   Add a whole slew of pegs—four black ones, two blue 3M pegs, and two cross connectors.

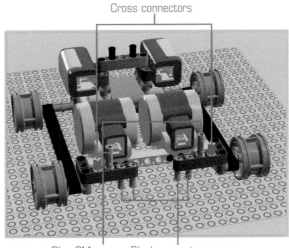

Cross connectors

Blue 3M peg   Black connector pegs

STEP 36   Pop on a 7M beam underneath the angle beams.

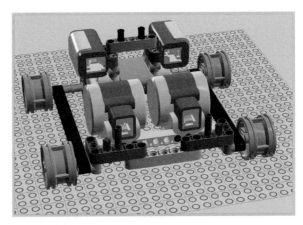

STEP 37   Add two more 3M pegs.

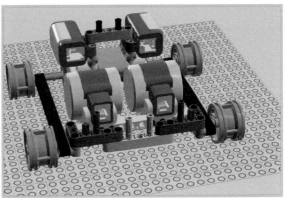

STEP 38   Throw on two 3M beams.

STEP 39   Add two 2M beams, one on each side. These have a regular peg hole as well as a cross hole.

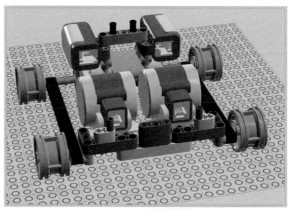

STEP 40   Add a 3M peg and push through so only one-third sticks out.

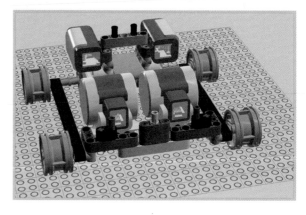

STEP 41   Add a 13M beam. You definitely want to plug in your motor wires before you perform this step, however!

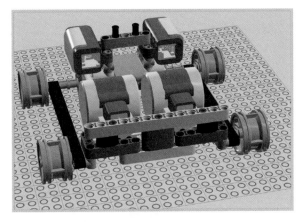

STEP 42   Add two pegs.

STEP 43   Add a 9M beam.

STEP 44 Connect two 3M beams with connectors.

STEP 45 Add the NXT brick!

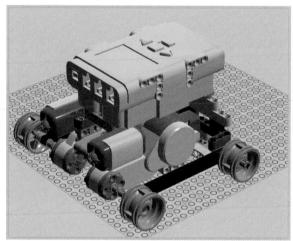

STEP 46 Finally, throw on that bumper you worked on at the beginning! You insert the bumper's 3M cross axles (the portions that are still visible) into the holes on the front of each Touch sensor.

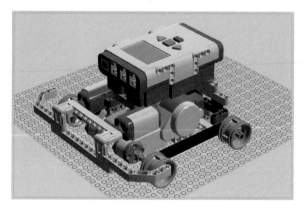

That's it! You're done!

A Note About Tank Treads

You may have noticed that the step-by-steps don't show the tank treads in place (see Figure 9.3). Don't worry! This is an unfortunate consequence of the CAD program I used to create the models. The treads pop on very readily; if they don't, just give the rubber a tug to stretch it out a bit.

FIGURE 9.3   These Mindstorms treads turn any car into a tank!

Programming the Rebounder

Finally, let's program the Rebounder to roll away from walls and other obstructions. I created a simple program (see Figure 9.4) that pulls the Rebounder away from something it hits and sends it off in another direction.

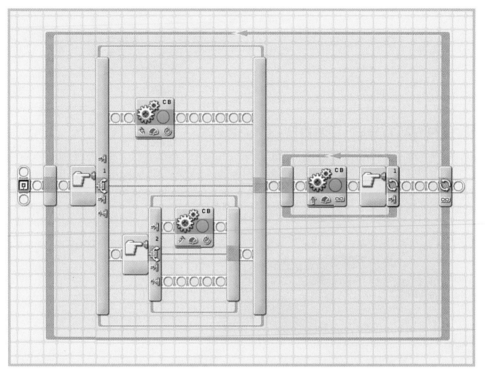

FIGURE 9.4   Program the Rebounder to react to its surroundings!

Let's get started with the Rebounder's program!

STEP 1 Open the Mindstorms software and create a new program.

STEP 2 Drag a Loop block out of the Common palette. It should be fine with its default settings. It should be configured to run Forever.

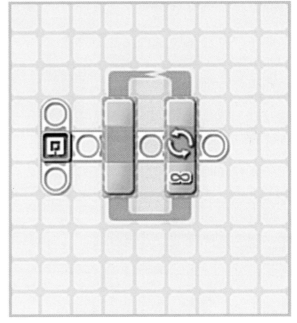

STEP 3 Drag a Switch block out of the Common palette and drop it into the Loop. The default settings of "Touch Sensor," "Port 1," and "Pressed" are fine.

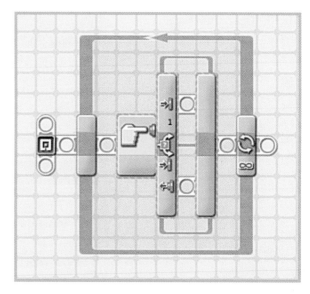

STEP 4 The Switch has two paths:

■ One for "button pressed" (the top sequence beam)

■ One for "button not pressed" (bottom)

For the former, drop a Move block into the top beam. Set Direction to Reverse (the downward-pointing arrow), Duration to 10 rotations, and Next Action to Coast. Pull the Steering tab all the way to the left. Power is fine at the default 75%.

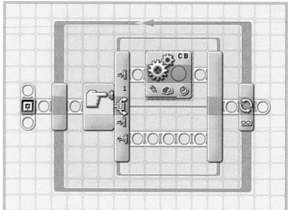

STEP 5   In the bottom sequence beam in the Switch, we're going to drop another Switch! Keep the settings identical, but go with Port 2 for the Touch sensor.

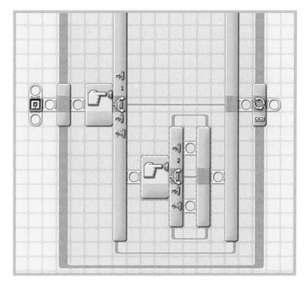

STEP 6   Add a Move block to the top sequence beam of the new Switch added in Step 5. Use the same settings as in Step 4 except pull the Direction tab all the way to the right. So basically, the robot looks for input from one Touch sensor, and if it doesn't detect anything, it looks at the other one.

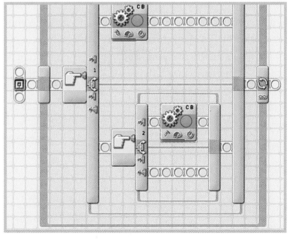

STEP 7   Drop another Loop inside the original Loop block but after the first Switch block. Set Control to Sensor, selecting the Touch Sensor on Port 1.

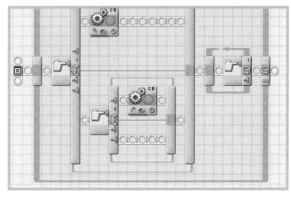

STEP 8 Inside this Loop, drop in a Move block with default settings except a Duration of Unlimited. This causes the robot to drive forward until it hits something, at which point it checks its sensors and decides what direction to turn. You're done! Your program should look like the one previously shown in Figure 9.4. Download the program to the Rebounder's NXT brick and start rolling!

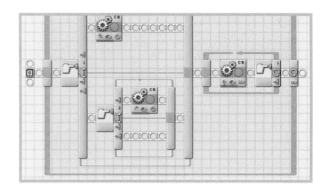

The Next Chapter

In Chapter 10, "Expanding on Mindstorms," we describe how to expand on LEGO Mindstorms. How do you get more LEGO parts? Where do other builders post their model designs and NXT-G programs? There's a whole ecosystem of Mindstorms fans whose work can inspire and educate you, and I give you a guided tour.

Expanding on Mindstorms

You've tackled programming, you've built three models, you've begun to explore the intricacies of the LEGO ecosystem. What else is there to do? Let's find out!

Read Blogs

Perhaps the best way to learn about all the possibilities of Mindstorms—other than by building—is to look at what other people have created. You'll see posts showing off the awesome creations of other builders, and reading about their work will inspire you to create your own. Let's look at a couple of good blogs.

The NXT STEP

thenxtstep.blogspot.com

The NXT STEP is the premier independent Mindstorms blog. It covers robots, reviews third-party products, and serves up robotics news to readers (see Figure 10.1). It's very low key and great for beginners, while supplying tons of information.

FIGURE 10.1 The NXT STEP blog serves up the best in independent Mindstorms news.

Mindstorms

mindstorms.lego.com

The official LEGO Mindstorms site (see Figure 10.2) is the central repository of information on the product. Message boards, technical support, a blog, plus downloadable resources are available for readers. Users can even share model designs, programs, and ideas with each other. It's the mother lode!

FIGURE 10.2   Literally the home of Mindstorms!

Design Virtual Models

You don't need to have a lot of parts in the real world if you're willing to work virtually through a computer-aided design (CAD) program such as LEGO Digital Designer (LDD), shown in Figure 10.3.

The way it works is that you're given a palette of LEGO parts—pretty much every single one, no matter how rare—and you drag them into the build area. The parts fit together just as you'd expect, and you can rotate the virtual model so you can see it from every angle, just as you could with a real model.

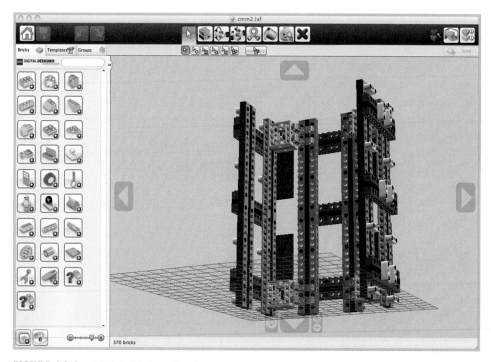

FIGURE 10.3 LEGO Digital Designer enables you to build any model you can imagine.

TIP

You've Already Seen LDD

You've already seen how cool LEGO Digital Designer can be. I used it to create many of the images in this book, showing step-by-step placement of each part.

Best of all, you can download LDD free! As long as you have a reasonably modern Mac or Windows computer, you'll be fine.

Attend Gatherings

Perhaps the best way to experience other builders' models is to see them in person. You get a better look at how others solved the problems they encountered along the way, and even better, you can talk to the builders directly! Next, let's look at some options for interacting with other Mindstorms fans in person.

LUGs

LEGO Users' Groups, or LUGs, are a great way to meet local LEGO builders (see Figure 10.4). LUGs are groups of fans who meet periodically to share models and talk about their hobby. Although a lot of groups focus on classic LEGO bricks, you'll always find Mindstorms fans among their number.

FIGURE 10.4   BayLUG is an organization for Bay Area LEGO fans.

Here are some of examples, but you'll definitely want to check out what's going on near you:

- BayLUG (San Francisco Bay Area)—baylug.org
- NCLUG (North Carolina)—nclug.us
- NELUG (New England)—nelug.org
- ToroLUG (Toronto)—torolug.com
- TwinLUG (The Twin Cities)—twinlug.com

Conventions

Numerous LEGO conventions are held around the world, including major ones such as BrickWorld in Chicago, featuring the work of hundreds of builders and featuring talks, vendor booths, official LEGO Group announcements, and other attractions. Even if you're not there to attend panels or show off your own work, LEGO conventions have "open days" (see Figure 10.5) where thousands of casual fans can check out the works of these expert builders.

FIGURE 10.5   Attendees at the BrickFair LEGO convention check out other fans' models. Credit: Joe Meno.

Here are some of the bigger conventions:

- **BrickCon**—brickcon.org—Held in Seattle every October.
- **BrickFair**—brickfair.com—Held three times a year in Birmingham, AL; Chantilly, VA; and Manchester, NH, BrickFair is believed to be one of the largest LEGO events in the world.
- **BrickWorld**—brickworld.us—Held in Chicago every June.

FIRST LEGO League

www.firstLEGOleague.org

One of the coolest learning opportunities for young LEGO aficionados is FIRST (see Figure 10.6), which stands for "For Inspiration and Recognition of Science and Technology." FIRST is a robotics competition that has expanded to include LEGO models in what is called FIRST LEGO League (FLL).

FIGURE 10.6   Kids work on their models while parents, judges, and other onlookers gather around. Credit: Grant Hutchinson.

The way it works is that an adult mentor helps a group of 9- to 14-year-old boys and girls build robots that compete in a variety of contests. The kids do everything from designing and building the robot to operating it during the competition. Examples of events include solving mazes and collecting objects scattered around the play surface. The mission is surprisingly difficult and takes an entire team to accomplish.

NOTE

Kid Friendly

FLL is great for kids because it fosters engineering know-how, encourages teamwork, and best of all, gives kids who can't afford a Mindstorms set the opportunity to play with this amazing learning tool.

Read *BrickJournal*

The LEGO phenomenon is big enough to support an actual print publication: *BrickJournal* (brickjournal.com)! The full-color and glossy magazine, which is available in major bookstores or online, covers masterful models, interviews builders and LEGO Group officials, and immerses readers in the AFOL (Adult Fan of LEGO) community (see Figure 10.7). Although primarily focusing on classic LEGO bricks, the publication does cover major Mindstorms news.

FIGURE 10.7   The gold standard for LEGO fan periodicals, *BrickJournal*.

Expand Your Collection

You've doubtlessly discovered by now that the Mindstorms set doesn't represent the entirety of LEGO's product line. The fact is, not only is it easy to expand on Mindstorms, but doing so is almost a necessity if you want to exploit the set to its full potential. Following are some great resources for buying additional parts.

Bricklink

bricklink.com

BrickLink (see Figure 10.8) is the equivalent to eBay for LEGO sellers. Need a particular part? You can almost certainly find it on BrickLink. At the time of this writing, the site has more than 150 million parts listed for sale.

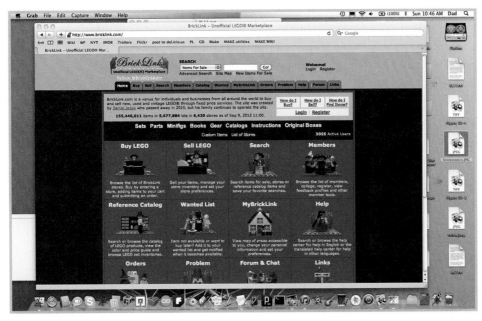

FIGURE 10.8   BrickLink's site, home of literally millions of bricks for sale!

Pick a Brick

shop.LEGO.com/en-US/Pick-A-Brick-ByTheme

LEGO's official resource for selling loose bricks, Pick a Brick consists of an online store as well as brick-and-mortar shops where hundreds of parts in many colors are available for purchase.

LEGO Education

legoeducation.us

Because schools use a lot of LEGO products, the company has created LEGO Education to serve this market with bulk packs of parts and special sets not available on the main website.

Third-Party Brick Makers

In Chapter 7, "Know Your Sensors," we covered a few of the more interesting third-party sensors available on the market. However, these parts are by no means all that is out there. Following are some examples of products that independent companies have created to support Mindstorms fans.

Omni Wheels

rotacaster.com.au

The wheels you get with Mindstorms aren't exactly robust. In fact, they're rather basic. One alternative is the omni wheel (see Figure 10.9) sold by Rotacaster. These wheels feature smaller rollers along their diameter, enabling the wheels to move sideways if needed. They can be pretty pricy, upwards of $10 each, but they're so much sweeter than those default wheels.

FIGURE 10.9   Omni wheels enable you to go sideways as well as back and forth. Credit: Rotacaster Wheel Limited.

Bricktronics

wayneandlayne.com/bricktronics

Produced by electronic kit seller Wayne & Layne, Bricktronics (see Figure 10.10) mates the coolness of Mindstorms with the versatility of the Arduino microcontroller. Arduinos are inexpensive ($35) but powerful microcontrollers designed to be easy to program, even for absolute newbies. Wayne & Layne's Bricktronics Shield allows you to use the Mindstorms wires, sensors, and beams while omitting the NXT brick.

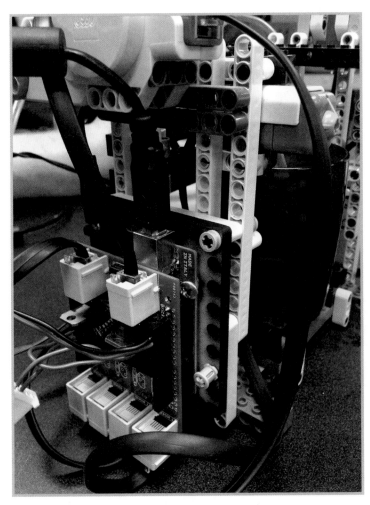

FIGURE 10.10   Forget the NXT brick; an Arduino can tell your robot what to do.

Tetrix

tetrixrobotics.com

LEGO beams may be strong enough for most models, but sooner or later you'll want something stronger. You'll have to turn to Tetrix Robotics (see Figure 10.11), which offers kits of metal beams, gears, and connectors, enabling you to create an all-metal robot that can be controlled by an NXT brick and uses LEGO sensors.

FIGURE 10.11   Just like Mindstorms, only metal. Credit: Pitsco Education.

Print Your Own

One of the more radical ideas that has come along is the possibility of actually *printing* beams (see Figure 10.12) and other parts using machines called *3D printers*. These gadgets lay down layers of melted plastic. When the extrusions cool down, the parts are (theoretically) just as useful as parts molded by LEGO! Right now, the technology isn't any threat to the LEGO Group's profit margin, but it does offer some intriguing possibilities for builders who can't buy a certain part or the part they need simply doesn't exist.

FIGURE 10.12   Real LEGO beams mate with printed ones. Credit: Brian Jepson.

For example, if you need a 16M beam, guess what? LEGO makes them only in odd-numbered sizes. But a builder with access to one of these machines need only design and print that 16M beam to get one. It's not exactly a cinch, and those printers cost upward of a thousand dollars, but the possibilities are fun to think about.

The Universal Connector Kit (see Figure 10.13) is a radical twist on the already-crazy idea of printing your own bricks. The kit consists of LEGO parts combined with connectors from other building sets such as Tinker Toys, Zoobs, K'Nex, Lincoln Logs, and many more. Simply put, this connector kit—which the LEGO Group would *never* produce on its own—enables builders to combine their favorite building sets in one creation.

See fffff.at/free-universal-construction-kit/.

FIGURE 10.13   The Universal Connector Kit enables you to connect every building set to every building set! Credit: Golan Levin (F.A.T. Lab) and Shawn Sims (Sy-Lab).

Glossary

The following terms are ones mentioned in the book, gathered for your convenience. There's a wide vocabulary used by LEGO robotics fans, ranging from programming terms unique to Mindstorms to concepts common across all flavors of robotics.

A

Analog: An electronic signal that delivers information through variances in frequency and amplitude in a continuous stream, as opposed to a digital signal, which returns the result as a series of numbers.

Angle Beam: A Technic beam that has one or two bends along its length.

Angle Connector: Technic connector that comes in different angles, allowing parts to be connected in ways other than the usual (and boring!) 90 degrees.

Autonomous: A robot's capacity to be controlled by a microcontroller rather than relying on human input.

B

Block Import and Export Wizard: The Mindstorms software tool that enables you to add other people's custom blocks and to share ones you created.

Bluetooth: A wireless communication protocol used for computer peripherals and mobile phone accessories. The Mindstorms NXT brick uses Bluetooth to connect to computers and other NXT bricks.

Bluetooth Connection Block: A NXT-G programming block that establishes a Bluetooth connection with another NXT brick.

Bush: A Technic connector used to secure the end of a cross axle.

C

CAD: Computer-Aided Design. The computer-generated drawings used for the step-by-step procedures in this book are examples of CAD.

Calibrate Block: A Mindstorms programming block that can calibrate the various sensors.

Color Lamp Block: A Mindstorms programming block that turns the color sensor into a lamp that emits either red, blue, or green light.

Color Sensor Block: A block that detects the color of an object by shining a lamp on it and measuring the light reflected back.

Comb Wheel: Also known as a cam, an oval-shaped wheel used to translate rotational motion (such as produced by a motor's axle) into linear (back and forth) motion.

Common Palette: A block palette in NXT-G that includes the most commonly utilized blocks.

Compare Block: A block used to compare two different numbers and outputs a "true" or "false" depending on the settings you choose in the configuration panel.

Complete Palette: The NXT-G block palette that contains every block available to the user.

Configuration Panel: The part of the screen displaying the settings you chose for a particular block.

Conical Gear: A beveled gear that can be paired with another at an angle, connecting two perpendicular axles.

Constant Block: A block used to store a fixed number or string of text, which remains the same throughout the operation of the robot.

Cross Axle: Technic rod used both as an axle and support in a Mindstorms project.

Cross Axle Extension: Tube for connecting two cross axles to make a longer one.

Cross Block: 3M LEGO beam with Technic holes at 90-degree angles.

Cross Connector: A 2M Technic peg with a cross axle on one end.

Custom Palette: The block palette in the Mindstorms software that holds user-created blocks, including those created by other users.

Cycle: One circuit of a NXT-G program.

D

Data Hub: The drop-down interface menu built into nearly every block, allowing the transfer of various types of information between blocks, with the help of data wires.

Digital: The type of data delivered as a series of numbers rather than a continuous stream of data.

Display Block: The NXT-G block that controls the display of information on the NXT brick's screen.

Driven Gear: The final gear in a train of gears.

E

Encoder: The slotted disc inside a LEGO servo that sends position information (that is, feedback) back to the NXT brick.

F

Feedback: Position data that a Mindstorms servo motor sends back to the NXT brick.

File Access Block: The block used to read and write text or numerical information onto files stored on the NXT.

Flexible Axle Damper 2M Connector: Rare in that it's not made of LEGO's usual plastic, a rubber part used to connect two LEGO elements in a flexible manner.

G

Gear: A toothed wheel, typically paired with other gears, to transfer motion to various parts of a robot.

Gear Ratio: A value that represents the ratio between the speed of the motor and the speed of the driven gear.

Gear Train: A series of gears with their teeth interlocked.

H

Hexapod: A six-legged robot.

Hub: The part of a motor that turns; the Mindstorms motors have an orange hub.

I

Idler Gear: An unpowered gear in a gear train, used mainly to transfer motion to the next gear in the train.

K

Keep Alive Block: The block that keeps the NXT brick from going into sleep mode. (Should probably be called the "Keep Awake" block!)

Knob Gear: A gear sporting large knobs instead of the usual teeth.

L

Light Sensor Block: The NXT-G block for controlling the light-sensing functions of Mindstorms' Color Sensor.

Linear Actuator: An assembly that converts rotary motion (for example, a motor) into linear motion (moving in a line). It's used for shoving and pulling objects.

Logic Block: A block that delivers a logic result when triggered. In NXT-G terminology, "logic" describes something that can be either Yes or No.

Loop Block: A block that repeats a series of actions until the NXT brick shuts down, the program is canceled, or a specific trigger (for example, a button press) is performed.

M

Math Block: A block that applies a mathematical function, such as addition or division, to two numbers.

Microprocessor: The mini computer inside the NXT brick that controls its electronic functions.

Module ("M"): The standard LEGO measurement, the M, which equates to 8mm.

Motor Block: A block that controls a single motor.

Move Block: A block that controls two motors and serves as a default motor control block for rolling robots. A more complicated version of the Motor block.

My Block: A custom block created by Mindstorms' My Block function that takes a segment of code and transforms it into a single block.

My Portal: An option on the Mindstorms menu that connects you to LEGO community resources.

N

Number to Text Block: A block that converts number data into text data.

NXT Buttons Block: A block that allows the program to interact with the buttons on the NXT brick.

NXT Controller: The Mindstorms menu used to access the NXT brick.

NXT Intelligent Brick: Another, fancier name for the NXT microcontroller brick, or "brick" as I refer to it throughout this book.

NXT-G: The default Mindstorms programming language.

P

Peg: The primary connector used in a Mindstorms set.

Potentiometer: A device that consists of a knob or slider that can gradually change the amount of current allowed to get through, similar to a light dimmer.

Proximity Sensor: A sensor that can tell when an object gets close.

R

RAM: Random Access Memory; a computer's memory. RAM stores data on a chip: the NXT 2.0 brick holds 64,000 kilobits of information. However, when the NXT brick powers down, the data stored in RAM is lost.

Random Block: A NXT-G block that generates a random number between 0 and 100, or else a smaller subset of that range: for instance, 5–25 or 1–50.

Range Block: A block that accepts a number and determines whether it falls within or outside a specific range. If a sound or light sensor returns a reading of 53, for instance, and the Range block is set to trigger an action if the result is between 55 and 75, then that action would not take place.

Receive Message Block: A NXT-G block that accepts wireless messages from other NXT bricks, enabling the robot to receive data via Bluetooth.

Record/Play Block: The block that records the motion of a specific motor and plays it back, making the motor exactly replay its rotations.

Remote Control: The menu option in the Mindstorms software that lets users control a robot from their computer.

Rotation Sensor Block: The block that keeps track of degrees the motor's hub turns, or complete rotations, informing the NXT brick of how far the robot has traveled. You can turn a motor into a sensor with this NXT-G block!

S

Send Message Block: A Bluetooth control block in NXT-G that controls the sending of data from one NXT brick to another wirelessly.

Sequence Beams: The pretend LEGO beams that show the order of a program's operations in NXT-G.

Servo Motor: The motor featuring a feedback mechanism that informs the microcontroller (in the case of Mindstorms, the NXT brick) of its position.

Snake Robot: A robot shaped like a snake that moves approximately like a snake.

Sound Block: A block that plays a sound on the NXT brick's speaker. Users can choose between a simple tone and one of the sound files currently stored on the NXT.

Sound Sensor Block: The NXT-G block that interfaces with the sound sensor, which triggers actions in the program when it detects noises above a certain level. The sound sensor doesn't ship with the Mindstorms 2.0 set. You can, however, purchase one separately here: http://www.legoeducation.us/eng/product/nxt_sound_sensor/2227

Stop Block: A NXT-G programming block that ends the program immediately.

Switch Block: A programming block that splits the program into multiple paths and follows one path or another on the basis of specific sensor input. For example, the robot could roll to the left if there's an obstruction on the right, and vice versa.

T

Technic: The LEGO product line from which the beams and connectors featured in the Mindstorms set hail. It was developed to allow robot builders to create stronger models, as compared with the classic LEGO brick.

Text Block: A block that combines up to three pieces of text into a single output. The text pieces can be either sent from other blocks or typed in manually.

Timer Block: A block that keeps track of time and can trigger an action when a certain period of time has taken place.

Touch Sensor Block: The Mindstorms touch sensor is essentially a button, which, when pressed, sends a signal to the NXT brick, returning either a press-and-hold, a press-and-release, or a bumped action.

U

Ultrasonic Sensor Block: A sensor that detects obstacles by emitting a stream of inaudible sound that bounces back from the obstruction and is picked up by the sensor's microphone, similar to the way a bat can "see" in the dark.

V

Variable Block: A block that stores specific values for the use of other blocks. For instance, the robot could determine with its light sensor whether it's night or day; that status is stored in the Variable block.

W

Wait Block: A block that is just what its name sounds like: It pauses the program to wait for a timer to elapse or a sensor to be triggered.

Wires (Mindstorms): In Mindstorms parlance, these are black cables that look like phone cords. They can be used for either sensors or motors, and are reversible.

Wires (NXT-G): Also known as data wires, NXT-G wires control the flow of data from one block to another.

Index

Numbers

1/2 connector pegs, 16

2M axle pegs, 14

3D printing, 192-193

3M connector pegs, 12-13

3M cross axles, Backscratcher Bot project, 43

5M beams, Backscratcher Bot project, 47

5M cross axles, Backscratcher Bot project, 44

7M beams, Backscratcher Bot project, 46

15M beams, Backscratcher Bot project, 46

A

action figures (minifigs), 121

additional parts, finding
  3D printing, 192-193
  BrickLink website, 189
  LEGO Education website, 189
  Pick a Brick website, 189
  third-party brick makers
    Bricktronics website, 190

Omni Wheels website, 190

Tetrix Robotics website, 191

Alpha Rex robot, building, 7-8

angle beams, 24-25
  Backscratcher Bot project, 45
  stronger models, building, 124

angle elements, 26

arrow buttons (NXT bricks), 59

artwork, LEGO Mindstorms NXT 2.0 box, 9

axles
  2M axle pegs, 14
  3M cross axles, Back-scratcher Bot project, 43
  5M cross axles, Back-scratcher Bot project, 44
  cross-axle connectors, 27
  cross axles, 17
    building stronger models, 125-126
    cross axles with end stops, 18-19
  rubber axle connectors, 28

B

Backscratcher Bot, building, 37
  batteries, 38-39
  NXT bricks
    adding batteries to, 38-39
    connections, 93
    programming, 49-55, 92-93
    required parts list, 40-42
    running, 56
    step-by-step build instructions, 42-48

balls, 21

Barometric sensors, 140

batteries and NXT bricks, 38-39, 58, 77

BayLUG website, 185

beams
  5M beams, Backscratcher Bot project, 47
  7M beams, Backscratcher Bot project, 46
  15M beams, Backscratcher Bot project, 46
  angle beams, 24-25
    Backscratcher Bot project, 45

building stronger
models, 124
beams with pegs, 22
M numbers, 23
Bionicle teeth, 21, 43
blocks
Bluetooth blocks, 152
Constant blocks, 149
creating, 154-156
cross blocks, 27-28
Backscratcher Bot
project, 45
building stronger
models, 124-125
Display blocks, 152
Keep Alive blocks,
150-151
Light Sensor blocks, 151
Logic blocks, 153
Loop blocks, 88,
177-179
Move blocks, 89,
179-180
NXT-G blocks
Color Sensor
blocks, 87
data hubs, 86
defining, 85
Loop blocks, 88
Move blocks, 89
NXT Buttons
blocks, 89
programming
blocks, 85
Switch blocks, 90
Ultrasonic Sensor
blocks, 91
Wait blocks, 91
pallettes, 87
Random blocks, 150

Receive Message blocks
(Bluetooth blocks), 152
Rotation Sensor
blocks, 151
Send Message blocks
(Bluetooth blocks), 152
Switch blocks, 90, 178
Variable blocks, 148-149
blogs (LEGO-related), 181
Bluetooth
Bluetooth blocks, 152
Bluetooth menu (NXT
bricks), 73-77
connections,
establishing, 77
NXT bricks, 58
box (LEGO Mindstorms
NXT 2.0)
artwork, 9
creative licensing and, 7
opening, 9
varying part quantities
in, 11
BrickCon, 186
BrickFair, 186
BrickJournal magazine, 188
BrickLink website, 189
brick makers (third-party)
Bricktronics website, 190
Omni Wheels
website, 190
Tetrix Robotics
website, 191
bricks
chassis bricks, building
stronger models, 128
NXT bricks, 35, 57
adding batteries to,
38-39
arrow buttons, 59

Backscratcher Bot
project, 48, 93
batteries, 58, 77
Bluetooth, 58,
73-77, 152
connector holes,
61-62
crashes, 78
Display blocks
and, 152
Enter (orange) but-
ton, 59
firmware updates,
78-79
go back (gray)
button, 59
gray (go back)
button, 59
Keep Alive blocks and,
150-151
My Files menu, 63-66
naming, 58
NXT Datalog
menu, 70
NXT Program
menu, 66
NXT Version item, 72
orange (Enter)
button, 59
ports, 59-60
powering, 77-78
reset button, 62-63
resetting, 78
Running icon, 58
Settings menu, 70-73
shutoff switch, 77-78
sleep feature, 72, 77
Try Me menu, 67
USB, 58
View menu, 68-70

System bricks
  building stronger
    models, 126-127
  original LEGO system
    bricks, 30
  Technic bricks, 126-127
Bricktronics website, 190
BrickWorld, 186
broken connections, gray
  wires and, 146
build competitions,
  186-187
build instructions (step-by-
  step)
  Backscratcher Bot, 42-48
  Clothesline Cruiser, 97,
    102, 113, 119
  Rebounder robot, 159-
    161, 166, 170, 173-175
bumps (Touch sensors), 132
bushings, 16
  Backscratcher Bot proj-
    ect, 44
  connector pegs with
    bushings, 13

C

CAD (computer-aided
  design) programs, LDD,
  183-184
calibrating
  blocks, 137
  sensors, 136
cams, 28
car parts, 25
chassis bricks, building
  stronger models, 128
Clothesline Cruiser
  parts needed, 96
  programming, 119

setting up, 120
step-by-step build
  instructions, 97, 102,
  113, 119
Touch sensors, 95
ultrasonic sensors, 95
uses for, 121
color
  Color Sensor blocks
    (NXT-G), 87
  color sensors, 31-33
    availability of, 134
    color lamps, 133
  LEGO Mindstorm NXT
    2.0 parts, 10
combination parts, building
  stronger models, 124-125
Common Pallette, 87
compass sensors, 138
competitions (build),
  186-187
Complete Pallette, 87
Configuration panel (Mindstorms work area), 87
connections
  broken connections and
    gray wires, 146
  multiple connections,
    building stronger
    models, 123
connector holes, NXT
  bricks, 61-62
connector pegs, 11
  1/2 connector pegs, 16
  3M connector pegs,
    12-13
  Backscratcher Bot proj-
    ect, 45-46
  connector pegs with
    bushings, 13

connector pegs with
  towballs, 15
Constant blocks, 149
conventions/gatherings
  BrickCon, 186
  BrickFair, 186
  BrickWorld, 186
  LUGs, 185
corners (reinforced), building strong models, 124
crashes (NXT bricks), resetting from, 78
Create My Block, 154
cross axles, 17, 27
  3M cross axles, Back-
    scratcher Bot
    project, 43
  5M cross axles, Back-
    scratcher Bot
    project, 44
  cross axles with end
    stops, 18-19
  stronger models, build-
    ing, 125-126
cross blocks, 27-28
  Backscratcher Bot proj-
    ect, 45
  stronger models,
    building, 124-125
cross connectors, 14
cross holes, 14
Custom Pallette, 87

D

data hubs, 86
datalog files (NXT brick My
  Files menu), 66
data wires
  connecting, 147
  gray wires, 146

green wires, 144
orange wires, 145-146
physical wires vs., 143
yellow wires, 145
Delete Files option (NXT bricks), 73
deleting
files
replacing deleted files, 73
Settings menu (NXT bricks), 71
preflight files, 84
sound files, 65
diagnostics, View menu (NXT bricks), 68
dIMU (Inertial Motion) sensor, 140
Display blocks, 152
downloading programming, Backscratcher Bot, 94

E - F

end stops (cross axles with), 18-19
Enter (orange) button (NXT bricks), 59

figures (minifigs), 121
files (NXT brick My Files menu)
datalog files, 66
NXT files, 65
software files, 64
sound files, 65-66
finding
additional parts
3D printing, 192-193
BrickLink website, 189
Bricktronics website, 190

LEGO Education website, 189
Omni Wheels website, 190
Pick a Brick website, 189
Tetrix Robotics website, 191
Universal Connector Kit, 193
Technic bricks online, 127
firmware updates, NXT bricks, 78-79
Flex sensors, 139
FLL (FIRST LEGO League), 186-187
further reading
BrickJournal magazine, 188
Mindstorms website, 182
NXT STEP blog, 181

G

gatherings/conventions
BrickCon, 186
BrickFair, 186
BrickWorld, 186
LUGs, 185
gears, 20
gray (go back) button (NXT bricks), 59
gray wires, 146
green wires, 144

H

half bushings, 16
help
BrickJournal magazine, 188

LEGO Mindstorms User Guide, 10
Mindstorms User Guide, 49
Mindstorms website, 182
NXT STEP blog, 181
programming, trouble-shooting, 94
Help window (Mindstorms work area), 87
holes
connector holes (NXT bricks), 61-62
cross holes, 14
Technic holes, 14

I

Inertial Motion sensors, 140
Infrared (PIR) sensors, 138
installing NXT-G
Mac installations, 84-85
PC installations, 83-84
instructions (step-by-step)
Backscratcher Bot, 42-48
Clothesline Cruiser, 97, 102, 113, 119
Rebounder robot, 159-161, 166, 170, 173-175
interactive servo motors, 32

J - K - L

joiners (peg), 29

Keep Alive blocks, 150-151

launchers (ShooterBot), 28
LDD (LEGO Digital Designer), 183-184

legged robots, Touch sensors, 132

LEGO Education website, 189

LEGO Mindstorms NXT 2.0 box
artwork, 9
creative licensing and, 7
opening, 9
varying part quantities in, 11

LEGO Mindstorms User Guide, 10

length. *See* M numbers

Leopard (Mac OS 10.5), NXT-G installations, 85

liftarms, Backscratcher Bot project, 46

Light Sensor blocks, 151

Linux, NXT-G system requirements, 82

Logic blocks, 153

Logic data and green wires, 144

Loop blocks
NXT-G, 88
Rebounder robot, 177-179

LUGs (LEGO User Groups), 185

M

M numbers, 17
2M axle pegs, 14
3M connector pegs, 12-13
3M cross axles, Backscratcher Bot project, 43
5M cross axles, Backscratcher Bot project, 44

beams, 23
5M beams, Backscratcher Bot project, 47
7M beams, Backscratcher Bot project, 46
15M beams, Backscratcher Bot project, 46

cross axles
3M cross axles, Backscratcher Bot project, 43
5M cross axles, Backscratcher Bot project, 44

Macs, NXT-G installations
Leopard (10.5) installations, 85
preflight files, 84
Snow Leopard (10.6) installations, 85
system requirements, 82

magazines (LEGO-related)
BrickJournal, 188
ShooterBot, 28

Magnetic sensors, 139

menus (NXT bricks)
Bluetooth menu, 73-77
My Files menu, 63
datalog files, 66
NXT files, 65
software files, 64
sound files, 65-66
NXT Datalog menu, 70
NXT Program menu, 66
Settings menu, 70
Delete Files option, 73
Volume menu, 71
Try Me menu, 67

View menu, 68-70
Volume menu, 71

Mindstorms software
Clothesline Cruiser, programming, 119
Mindstorms User Guide, 49
Mindstorms website, 182
Rebounder robot, programming, 176-180
welcome screen, 82
work area, 86-87

minifigs, 121

models (virtual), LDD, 183-184

motion sensors (Inertial), 140

motors, 31
Backscratcher Bot project, 46
interactive servo motors, 32
NXT brick ports, 60
sensors as, 135

Move blocks
NXT-G, 89
Rebounder robot, 179, 180

multiple connections, building stronger models, 123

multiple pegs, building stronger models, 123

My Files menu (NXT bricks), 63
datalog files, 66
NXT files, 65
software files, 64
sound files, 65-66

My Portal (Mindstorms work area), 87

N

naming NXT bricks, 58

NCLUG website, 185

NELUG website, 185

Number data and yellow wires, 145

NXT bricks, 35, 57

  arrow buttons, 59

  Backscratcher Bot project, 48, 93

  batteries, 38-39, 58

  Bluetooth, 58, 73-77, 152

  connector holes, 61-62

  crashes, resetting from, 78

  Display blocks and, 152

  Enter (orange) button, 59

  firmware updates, 78-79

  gray (go back) button, 59

  Keep Alive blocks and, 150-151

  menus

    Bluetooth menu, 73-77

    My Files menu, 63-66

    NXT Datalog menu, 70

    NXT Program menu, 66

    Settings menu, 70-73

    Try Me menu, 67

    View menu, 68-70

  naming, 58

  NXT Version item, 72

  orange (Enter) button, 59

  ports, 59-60

powering

  batteries, 77

  shutoff switch, 77-78

  sleep feature, 77

reset button, 62-63

Running icon, 58

sleep feature, 72, 77

USB, 58

NXT Buttons blocks (NXT-G), 89

NXT controller (Mindstorms work area), 87

NXT Datalog menu (NXT bricks), 70

NXT files (NXT brick My Files menu), 65

NXT-G

  Backscratcher Bot, programming, 92

  blocks

    data hubs, 86

    defining, 85

  Clothesline Cruiser, programming, 119

  Color Sensor blocks, 87

  Linux system requirements, 82

  Loop blocks, 88

  Macs

    installing on, 84-85

    Leopard (10.5) installations, 85

    Snow Leopard (10.6) installations, 85

    system requirements, 82

  Move blocks, 89

  NXT Buttons blocks, 89

  PCs

    installing on, 83-84

    system requirements, 82

  programming blocks, 85

  Rebounder robot, programming, 176-180

  Switch blocks, 90

  Ultrasonic Sensor blocks, 91

  Wait blocks, 91

NXT Program menu (NXT bricks), 66

NXT STEP blog, 181

NXT Version item (NXT bricks), 72

O - P

Omni Wheels website, 190

orange (Enter) button (NXT bricks), 59

orange wires, 145-146

pallettes

  Common Pallette, 87

  Complete Pallette, 87

  Custom Pallette, 87

parts (additional), finding

  3D printing, 192-193

  BrickLink website, 189

  LEGO Education website, 189

  Pick a Brick website, 189

  third-party brick makers

    Bricktronics website, 190

    Omni Wheels website, 190

    Tetrix Robotics website, 191

PCs, NXT-G installations, 82-84

peg joiners, 29

pegs
2M axle pegs, 14
3M connector pegs, 12-13
beams with pegs, 22
connector pegs, 11
1/2 connector pegs, 16
3M connector pegs, 12-13
Backscratcher Bot project, 45-46
connector pegs with bushings, 13
connector pegs with towballs, 15
cross connectors, 14
multiple pegs, building stronger models, 123
permissions, NXT-G PC installations, 84
Pick a Brick website, 189
pins (Technic), 28
PIR (Passive Infrared) sensors, 138
ports (NXT bricks), 59-60
powering NXT bricks
batteries, 77
shutoff switch, 77-78
preflight files, deleting, 84
presses (Touch sensors), 132
printing (3D), 192-193
programming
Backscratcher Bot, 49-55
creating programming, 92
downloading programming, 94
NXT brick connections, 93
calibrating blocks, 137
Clothesline Cruiser, 119
Rebounder robot, 176-180
troubleshooting, 94
programming blocks (NXT-G), 85

R

Random blocks, 150
reading
BrickJournal magazine, 188
BrickLink website, 189
LEGO Mindstorms User Guide, 10
Mindstorms User Guide, 49
NXT STEP blog, 181
ShooterBot magazine, 28
Rebounder robot, 157
Loop blocks, 177-179
Move blocks, 179-180
parts needed, 158
programming, 176-180
step-by-step build instructions, 159-161, 166, 170, 173-175
Switch blocks, 178
tank treads, 175
Receive Message blocks (Bluetooth blocks), 152
rechargeable batteries, NXT bricks, 77
reinforced corners, building stronger models, 124
releases (Touch sensors), 132
replacing deleted files, 73
reset button (NXT bricks), 62-63
resetting NXT bricks, 78
rims and tires, 19
RoboCenter (Mindstorms work area), 87
Rotation Sensor blocks, 151
rubber axle connectors, 28
Running icon (NXT bricks), 58

S

Send Message blocks (Bluetooth blocks), 152
sensors, 131
Barometric sensors, 140
calibrating, 136-137
Color sensors, 31-33
availability of, 134
color lamps, 133
compass sensors, 138
dIMU, 140
Flex sensors, 139
Inertial Motion sensors, 140
Magnetic sensors, 139
motion (Inertial), 140
motors as, 135
NXT brick ports, 60
PIR sensors, 138
Sound sensors, 134
Touch sensors, 31-32, 95, 132
Ultrasonic sensors, 31, 95, 133
voltage sensors, 140
VoltMeter, 140
Wi-Fi sensors, 138
sequence beams (Mindstorms work area), 87
servo motors (interactive), 32

Settings menu (NXT bricks), 70
  Delete Files option, 73
  Volume menu, 71
ShooterBot magazine, 28, 49
shutoff switch (NXT bricks), 77-78
sleep feature (NXT bricks), 72, 77
Snow Leopard (Mac OS 10.6), NXT-G installations, 85
software files (NXT brick My Files menu), 64
sound files
  NXT brick My Files menu), 65-66
  Volume menu (NXT bricks), 71
Sound sensors, 134
starting point (Mindstorms work area), 87
steering links, 26
stronger models, building techniques
  angle beams, 124
  chassis bricks, 128
  combination parts, 124-125
  cross axles, 125-126
  cross blocks, 124-125
  multiple connections, 123
  multiple pegs, 123
  reinforced corners, 124
  System bricks, 126-127
  Technic bricks, 126-127
Switch blocks

NXT-G, 90
  Rebounder robot, 178
System bricks
  original LEGO system bricks, 30
  stronger models, building, 126-127
system requirements (NXT-G)
  Linux, 82
  Macs, 82
  PCs, 82

T

tank treads, 19, 175
Technic bricks
  stronger models, building, 126-127
  web resources, 127
Technic holes, 14
Technic pins, 28
teeth (Bionicle), 21, 43
Tetrix Robotics website, 191
Text data and orange wires, 145-146
third-party brick makers
  Bricktronics website, 190
  Omni Wheels website, 190
  Tetrix Robotics website, 191
tires and rims, 19
toolbar (Mindstorms work area), 87
tooth gears, 20
ToroLUG website, 185

Touch sensors, 31-32, 95, 132
towballs, connector pegs with, 15
treads, 19
troubleshooting programming, 94
Try Me menu (NXT bricks), 67
TwinLUG website, 185

U

ultrasonic sensor, 31
Ultrasonic Sensor blocks (NXT-G), 91
Ultrasonic sensors, 95, 133
Universal Connector Kit, 193
updating NXT brick firmware, 78-79
USB
  NXT brick ports, 59-60
  NXT bricks, 58
  USB cable, 31, 34
user profile (Mindstorms work area), 87

V

Variable blocks, 148-149
versioning, NXT Version item (NXT bricks), 72
View menu (NXT bricks), 68-70
virtual models, LDD, 183-184
voltage sensors, 140
VoltMeter, 140
Volume menu (NXT bricks), 71

W

Wait blocks (NXT-G), 91

web resources, Technic bricks, 127

websites
   BayLUG, 185
   BrickLink, 189
   Bricktronics, 190
   LEGO Education, 189
   NCLUG, 185
   NELUG, 185
   Omni Wheels, 190
   Pick a Brick, 189
   Tetrix Robotics, 191
   ToroLUG, 185
   TwinLUG, 185
   Universal Connector
      Kit, 193

welcome screen (Mind-storms software), 82

wheels, 19, 190

Wi-Fi sensors, 138

Windows. *See* PCs

wires, 31, 35
   Backscratcher Bot
      project, 48
   data wires, 143
      connecting, 147
      gray wires, 146
      green wires, 144
      orange wires, 145-146
      yellow wires, 145
   Mindstorms work
      area, 87
   physical wires, 143

work area (Mindstorms), 86-87

X - Y - Z

yellow wires, 145